D1335934

9030 00003 4551 9

Chrissy Freer is a food writer, qualified nutritionist, stylist and editor whose signature style is creating delicious recipes with a wholistic health focus. Her interest in ancient grains reflects her passion for creating nutritious, flavoursome and easy to prepare food that makes the most of natural ingredients. She has worked for numerous magazines, including *delicious.*, *australian good taste*, *Belle*, *Fresh*, *Recipes+*, *Healthy Food Guide*, *Super Food Ideas* and *Weight Watchers*, and on cookbooks including *Everyday* and *Holiday* by Bill Granger, the *Indulgence* series, *The Biggest Loser Family Cookbook* and many *Weight Watchers* titles. Chrissy lives with her daughter Harriet and partner Tim in Melbourne, Australia.

CHRISSY FREER

supergrains

eat your way to great health

MURDOCH BOOKS

Contents

Introduction

What are supergrains?

In today's society, the impact food can have on our health is becoming increasingly important. A superfood is a food that by nature is nutrient dense compared to its energy value. Supergrains are grains that are not only packed with nutrients, but have also been linked to many health benefits, which is why they are currently gaining worldwide interest.

Let's start by defining grains. These are the seeds of certain plants, where the hard inedible husk or outer layer is removed, leaving the edible grain (or berry). Unrefined (whole) grains contain all three parts of the grain: the bran, germ and endosperm. Supergrains (in their simplest forms) are all unrefined grains. Refined grains have the bran and germ removed, leaving just the endosperm which contains the least vitamins and minerals.

This book explores 12 supergrains: quinoa, amaranth, buckwheat, brown rice, chia, millet, oats, spelt, kamut, barley, farro and freekeh. We look at their history, nutritive and health benefits, what they are best used for and how to cook them. There are delicious recipes for each supergrain so you will have the confidence to incorporate them into your everyday life.

Although some of the supergrains are quite common, others you may not have heard of, despite the fact that they have been around for centuries. Spelt, for instance, was a staple grain in ancient civilisations such as Greece and Rome. One of the main reasons for their decline in popularity was the modernisation of farming methods, which produced more refined, highly processed foods. Compared to these new, refined foods, supergrains came to be considered by many cultures as peasant food, only eaten by the very poor or even fed to the animals.

The problem with replacing supergrains with refined grains, however, is that the latter are much lower in nutrients, healthy fats, antioxidants and phytonutrients. Although most of the grains currently consumed in first-world countries are highly refined, such as white rice, white wheat flour and ground cornmeal, there is renewed interest in supergrains for their health benefits, superior taste and suitability for restricted diets, such as gluten and wheat intolerances, vegetarians and vegans.

Another reason for the re-emergence of supergrains is a growing interest in making food more 'simplistic', by going back to its original source and subjecting it to minimal refining. Farmers' markets, the organic and biodynamic food movements, the demand for quality ingredients and the move towards 'whole' foods all indicate an increasing awareness and interest in what goes into food products and their natural health benefits.

As supergrains gain in popularity, their availability also improves. Quinoa, for example, was once considered a 'speciality' grain, only found in serious health food stores and on restaurant menus, but today it can be found in almost every major supermarket. Kamut and spelt are enjoying a renewed popularity, too, especially in the form of baked products such as bread. Chia is also receiving a lot of attention as it contains more omega 3 and dietary fibre than any other natural food.

The recipes in this book reflect my approach to health and eating — everything in moderation, with a focus on whole foods, such as supergrains. Eating responsible portion sizes means that occasional treats can be enjoyed (and, in fact, are a must!). Meals should be balanced, with one portion of protein, one portion of complex carbohydrate and two to three serves of vegetables. I hope these delicious recipes inspire you and contribute to your own healthy eating.

Health benefits

Controlling blood sugar levels

The glycaemic index (GI) is a measure of the speed with which individual carbohydrate foods are converted into blood sugar. The lower the GI of a food, the slower it releases glucose, resulting in sustained energy. High-GI foods, on the contrary, release glucose quickly and give a spike in blood sugar. As supergrains are complex carbohydrates, they generally have a low GI, providing slowly released energy that is extremely beneficial for everyone, particularly those with diabetes.

High fibre

Supergrains are an excellent source of dietary fibre, whether soluble, insoluble or both. Soluble fibre is water soluble, as the name suggests, and forms a thick gel-like substance. This gel helps keep the bowel healthy, reduces blood cholesterol and slows the release of sugars in the blood, which is very useful for those with diabetes. Insoluble fibre, on the other hand, stays intact even when in contact with water. Insoluble fibre helps in preventing constipation and irritable bowel syndrome, keeps your colon healthy and assists with heart health.

Reduced risk of obesity/weight management

Studies are beginning to show that supergrains can play an important role in reducing the risk of obesity when they are used as a tool for weight control. They are higher in fibre than highly processed grains, especially soluble fibre which can increase satiety so you feel full for longer and therefore eat less overall.

Prebiotic effect

Prebiotics are foods that stimulate the growth of healthy bacteria, helping to produce digestive enzymes. There have been recent studies on the prebiotic effects that supergrains can have in the digestive system, mainly due to the fermentation of the soluble fibre they contain.

Reduced risk of cardiovascular disease

The relationship between the consumption of supergrains and the reduced risk of cardiovascular disease is of great interest. Supergrains that are high in soluble fibre (such as oats and barley) have been proven to be effective in reducing blood cholesterol. Supergrains are also associated with a reduced risk of stroke, healthier carotid arteries and reduced blood pressure.

Reduced risk of certain cancers

Supergrains contain many natural compounds that protect cells from damage and therefore may reduce the risk of certain cancers. Insoluble fibre, in particular, is thought to reduce the risk of colon cancer by maintaining a healthy colon. Other compounds to be of benefit include antioxidants, phenols, lignans and saponins.

Cooking guide

Grain	Per 1 cup raw weight	Quantity of liquid required	Cooking time	Description	Yield (approx.)
Quinoa	200 g (7 oz)	500 ml (17 fl oz/ 2 cups)	12 minutes	Absorption method, grains turn translucent and uncurl, and have an *al dente* texture	555 g (1 lb 4 oz/ 3 cups)
Amaranth	210 g ($7\frac{1}{2}$ oz)	625 ml ($21\frac{1}{2}$ fl oz/ $2\frac{1}{2}$ cups)	20 minutes	Absorption method, grains become *al dente* with a slightly sticky texture	615 g (1 lb $5\frac{1}{2}$ oz/ $2\frac{1}{2}$ cups)
Buckwheat	200 g (7 oz)	500 ml (17 fl oz/ 2 cups)	10–12 minutes	Gentle boil, grains swell and keep some texture	510 g (1 lb $2\frac{1}{4}$ oz/ 3 cups)
Brown rice	200 g (7 oz)	375 ml (13 fl oz/ $1\frac{1}{2}$ cups) (absorption) 1.5 litres (52 fl oz/ 6 cups) (gentle boil)	25–30 minutes	Gentle boil or absorption method, grains swell and have a slightly chewy texture	585 g (1 lb $4\frac{1}{2}$ oz/ 3 cups)
Millet (hulled)	210 g ($7\frac{1}{2}$ oz)	500–625 ml (17–$21\frac{1}{2}$ fl oz/ 2–$2\frac{1}{2}$ cups)	20 minutes	Absorption method, grains tend to stick together slightly and then separate on cooling, and have an *al dente* texture	610 g (1 lb $5\frac{1}{4}$ oz/ $3\frac{1}{2}$ cups)
Oats (rolled)	95 g ($3\frac{1}{4}$ oz)	500 ml (17 fl oz/ 2 cups)	5–10 minutes	Absorption method, results in a creamy, soft porridge	470 g (1 lb $\frac{3}{4}$ oz/ 2 cups)
Kamut	200 g (7 oz)	1.5–2 litres (52–70 fl oz/ 6–8 cups)	45–50 minutes	Gentle boil, grains swell and have a slightly chewy texture	440 g ($15\frac{1}{2}$ oz/ $2\frac{1}{2}$ cups)
Spelt	200 g (7 oz)	1.5–2 litres (52–70 fl oz/ 6–8 cups)	45–50 minutes	Gentle boil, grains swell and have a slightly chewy texture	485 g (1 lb 1 oz/ $2\frac{1}{2}$ cups)
Pearl barley	200 g (7 oz)	1.5 litres (52 fl oz/ 6 cups)	30 minutes	Gentle boil, grains swell and have a slightly chewy texture	560 g (1 lb $4\frac{1}{4}$ oz/ $3\frac{1}{2}$ cups)
Farro (cracked)	175 g (6 oz)	1.5–2 litres (52–70 fl oz/ 6–8 cups)	15–20 minutes	Gentle boil, grains swell and keep some texture	425 g (15 oz/ $2\frac{1}{2}$ cups)
Freekeh (wholegrain)	200 g (7 oz)	1.5–2 litres (52–70 fl oz/ 6–8 cups)	45–50 minutes	Gentle boil, grains swell and have a slightly chewy texture	390 g ($13\frac{3}{4}$ oz/ $2\frac{1}{2}$ cups)

How to use supergrains

Grain	Flavour characteristics	Goes well with …	Use in …
Quinoa	Mild, nutty, slightly bitter	Extremely versatile. Great with Middle Eastern spices, Mediterranean flavours, fresh herbs, citrus and even Asian flavours	Salads, pilafs, stuffings for meat or vegetables, soups, stews, porridge and desserts
Amaranth	Herbaceous, grassy, sticky	Nuts, honey, chocolate and cinnamon, and Mexican flavours such as chilli, paprika and corn	Porridge, salads, soups and stews (when combined with other grains)
Buckwheat	Earthy, dark, slightly meaty	Foods that have natural sweetness and/or saltiness such as miso, roast beetroot, seafood and mushrooms, as well as creamy foods	Salads, risotto, pilafs, soups and muesli. Flour is used in pancakes and other baked goods
Brown rice	Nutty, slightly sweet, chewy	Other foods with nuttiness, saltiness or natural sweetness such as roasted vegetables, dried fruit and nuts, feta cheese and Asian flavours	Salads, risotto, pilafs, stuffings for meat or vegetables, stir-fries and rice pudding
Chia	Neutral, mild	Provides texture and will team well with almost anything, sweet or savoury. Delicious combined with oats, nuts, seeds, dried fruit and chocolate	Smoothies, dressings, baked goods and sprinkle over salads
Millet	Buttery, corn-like, mellow	Cheeses such as parmesan and feta, fresh herbs, corn, citrus and berries	Salads, pilafs, stuffings, porridge, soups, stews, desserts and to make a polenta-like dish
Oats	Creamy, slightly sweet, toasty	Cheese, nuts, chocolate, brown sugar and milk	Muesli, porridge, baked goods and coatings (instead of breadcrumbs)
Kamut	Buttery, nutty	Foods that are naturally slightly sweet or nutty, such as roasted vegetables and pine nuts	Salads, pilafs and stuffings for meat or vegetables. Flour is used in bread, pasta and baked goods
Spelt	Nutty, slightly earthy, chewy	Nuts, dried fruit and chocolate, and Mediterranean flavours such as basil, olives, tomato, cheese and eggplant	Salads, pilafs and stuffings for meat or vegetables. Flour is used in bread, pasta, baked goods and desserts
Pearl barley	Nutty, slightly chewy	Very versatile. Perfect with stronger flavours such as lamb, mushrooms, soy and miso	Salads, risotto, pilafs, stuffings, soups, stews and sweet puddings
Farro (cracked)	Nutty, mild, slightly chewy	Ideal with Italian flavours such as pancetta, herbs, parmesan and red wine	Salads, pilafs, risotto, stuffings, soups, stews and pasta
Freekeh (wholegrain)	Nutty, slightly herbaceous, green	Can take on quite robust flavours, such as cumin, paprika, pomegranate, lamb and mint	Salads, pilafs, stuffings for meat or vegetables, soups and stews

Quinoa

NUTRITIONAL INFORMATION
(per 185 g/6½ oz/1 cup cooked quinoa)

Energy: 929 kilojoules, 222 calories **Protein:** 8.1 g **Total fat:** 3.6 g
Saturated fat: 0 g **Carbohydrate:** 39.4 g **Sugars:** 0 g **Dietary fibre:** 5.2 g
Cholesterol: 0 mg **Folate:** 77.7 µg **Iron:** 2.8 mg **Magnesium:** 118 mg
Manganese: 1.2 mg **Phosphorus:** 281 mg **Sodium:** 13 mg

Quinoa (pronounced keen-wah) is an ancient grain of the Andes mountain range in South America, and was once the staple food of the Incas. It is enjoying a resurgence in popularity, and has gained recognition as a 'supergrain' due to its high levels of protein, balanced essential amino acid profile, rich nutrient content and gluten-free qualities. Its ease of preparation and versatility also make it stand out from some of the other grains.

Although it's considered and called a grain, quinoa is in fact a pseudo-grain, being the seed of a leafy plant distantly related to spinach, called *Chenopodium quinoa* (also known as goosefoot). Its history dates back some 5000 years, when it was second only to potatoes as a staple food of the Incas, who called it the 'mother seed'. They considered it sacred and believed it increased stamina, so their warriors ate it to keep them strong.

Today, quinoa has had a massive resurgence, as its unique nutritional qualities have been rediscovered. Quinoa is naturally gluten and wheat free, making it gentle on digestion and suitable for the ever-increasing number of people following a gluten-free diet. It is also an amazing food for vegetarians and vegans, not only because it is a rich source of protein, but also due to its balanced essential amino acid profile. It contains all the essential amino acids, making it a rare complete vegetable protein source. Of particular note is lysine, an amino acid that assists in tissue repair and growth.

In addition to protein, quinoa is rich in many nutrients, such as dietary fibre, and it is linked to reducing the risk of cardiovascular disease, certain cancers and type 2 diabetes. It is also high in manganese and is a good source of phosphorus, magnesium and folate.

Quinoa is available in various forms: as a whole grain, flakes and, less commonly, flour. Quinoa grains look like tiny flat discs. As the grain cooks the centre seed becomes translucent and soft, while the outside germ separates from the seed and looks like a white ring that retains an *al dente* texture. The grain can be found in various colours, but white, red and black are the most common. The red and black varieties have a slightly firmer texture once cooked and are a little stronger in flavour than the white variety.

One of the main attractions of quinoa is how easy and quick it is to cook. It does not require pre-soaking or toasting, and is ready in just 12 minutes. The grain is fabulous used in salads, as you would use rice, couscous or pasta. It can also be used to stuff vegetables and meat or added to soups, stews and curries, pilafs and more.

Quinoa flakes are even quicker to prepare, and can be used to make a delicious gluten-free porridge. They are fantastic added to baked goods such as muffins, breads and cookies, and make a crisp and golden coating when used as a substitute for breadcrumbs. Quinoa flour is harder to find, but it can be used in gluten-free baking, such as cakes and breads, or to make pasta. It does tend to have a slightly bitter taste, so is best combined with ingredients that have some natural sweetness.

To cook quinoa

Most of the quinoa available today has been pre-rinsed to remove the bitter 'saponin' coating around the grain, therefore it only requires a brief rinse before cooking.

Put the quinoa in a sieve and rinse for 30 seconds under cold running water. For every 200 g (7 oz/1 cup) of grain, add 500 ml (17 fl oz/2 cups) water, then bring to the boil over medium heat. Cover, reduce the heat to low and simmer for 12 minutes or until the water has evaporated and the grains are tender. (The red and black varieties of the grain may need a few extra minutes of cooking.) Remove from the heat and set aside to cool.

NOTE For a slightly nuttier flavour, toast the quinoa in a hot dry pan for 1–2 minutes, stirring, prior to cooking.

Breakfast quinoa with vanilla roasted plums

PREPARATION TIME: 10 MINUTES
COOKING TIME: 10 MINUTES
SERVES 4

8 (about 500 g/1 lb 2 oz) firm ripe plums, stones removed and cut into quarters
2 tablespoons caster (superfine) sugar
1 teaspoon vanilla bean paste
1 teaspoon lemon juice
145 g (5¼ oz/1⅓ cups) quinoa flakes
250 ml (9 fl oz/1 cup) milk
½ teaspoon ground cinnamon

Quinoa flakes make a fantastic gluten-free porridge that only takes minutes to cook and is packed with protein and carbohydrate. It's the perfect way to start the day and will keep you feeling full for a long time.

1 Preheat the oven to 180°C (350°F/Gas 4). Put the plums in a medium bowl, add the sugar, vanilla and lemon juice and toss to coat. Place the plums, cut side up, in a roasting pan and bake for 10 minutes or until tender. (The cooking time will vary depending on the ripeness of the plums.)

2 Meanwhile, put the quinoa flakes, milk, cinnamon and 500 ml (17 fl oz/2 cups) water in a large saucepan. Cook, stirring, over medium heat for 3–4 minutes or until thick and creamy.

3 Divide the quinoa porridge among 4 bowls, top with some roasted plums and spoon over any pan juices.

Banana, honey & walnut bread

PREPARATION TIME: 20 MINUTES
COOKING TIME: 45–50 MINUTES
SERVES 8

225 g (8 oz/1½ cups) plain (all-purpose) flour

1 teaspoon ground cinnamon

3 teaspoons baking powder

70 g (2½ oz/⅔ cup) quinoa flakes

60 g (2¼ oz/½ cup) walnut halves, coarsely chopped

175 g (6 oz/½ cup) honey, plus extra, to serve

2 eggs

80 ml (2½ fl oz/⅓ cup) sunflower oil

65 g (2¼ oz/¼ cup) sour cream

4 small (about 500 g/1 lb 2 oz) very ripe bananas, mashed

Fresh ricotta cheese, to serve

* *Quinoa flakes are ideal to use in baked goods where you would typically use rolled (porridge) oats.*

1 Preheat the oven to 180°C (350°F/Gas 4). Lightly brush a 9 x 19 cm (3½ x 7½ inch) loaf (bar) tin with oil and line the base with non-stick baking paper.

2 Sift the flour, cinnamon, baking powder and a pinch of salt into a large mixing bowl. Add the quinoa flakes and 40 g (1½ oz/⅓ cup) of the walnuts and stir to combine.

3 Combine the honey, eggs, oil, sour cream and banana in a separate bowl. Add to the dry ingredients and stir until just combined.

4 Spoon the mixture into the prepared tin and smooth the surface with the back of the spoon. Sprinkle the remaining walnuts over the top. Bake for 45–50 minutes or until golden and a skewer inserted into the centre comes out clean. Cover the top with foil if the loaf is browning too quickly.

5 Remove from the oven and set aside to cool for 10 minutes before turning out onto a wire rack. Serve warm or at room temperature with fresh ricotta and a drizzle of extra honey.

TIPS To make Banana & blueberry bread, omit the walnuts and stir through 155 g (5½ oz/1 cup) fresh (or frozen) blueberries at the end of step 3.

This banana bread is great to freeze and can then be toasted straight from the freezer for a quick breakfast. Just cut into slices, wrap in plastic wrap and freeze for up to 2 months.

gluten-free

Chickpea, tomato & quinoa soup with pesto toasts

PREPARATION TIME: 20 MINUTES
COOKING TIME: 35 MINUTES
SERVES 4

1 tablespoon olive oil
1 large red onion, finely chopped
1 large celery stalk, diced
2 garlic cloves, crushed
1 long red chilli, seeded and finely chopped
1 teaspoon ground cumin
2 teaspoons sweet paprika
1 tablespoon tomato paste (concentrated purée)
400 g (14 oz) tin chickpeas, drained and rinsed
65 g (2¼ oz/⅓ cup) quinoa (see tips), briefly rinsed
400 g (14 oz) tin chopped tomatoes
750 ml (26 fl oz/3 cups) vegetable or chicken stock (see tips)
8 slices sourdough baguette or gluten-free bread
Good-quality basil pesto (see tips), to serve

1 Heat the oil in a large saucepan over medium heat and cook the onion and celery, stirring, for 5–6 minutes or until soft. Add the garlic, chilli and spices and cook, stirring, for 1 minute, then add the tomato paste and cook, stirring, for 1 minute more.

2 Add the chickpeas, quinoa, tomatoes and stock and bring to the boil. Reduce the heat to low, partially cover and simmer for 25 minutes. Set aside to cool slightly.

3 Blend half the soup until smooth and return to the saucepan with the remaining soup. Place over medium heat, season with sea salt and freshly ground black pepper and warm through.

4 Meanwhile, toast the slices of bread until golden and then spread each with a little basil pesto. Serve the soup with the pesto toasts.

TIPS You can use either white or red quinoa in this recipe.

If you want this soup to be gluten-free make sure you use gluten-free stock and pesto.

This soup freezes well (without the pesto toasts). Simply place the cooled soup in airtight containers and freeze.

** This soup is perfect for vegetarians, as combining legumes (chickpeas) with a whole grain such as quinoa makes this a complete protein and therefore a balanced meal.*

Spiced quinoa, sweet potato, broccoli & cranberry salad

PREPARATION TIME: 20 MINUTES
COOKING TIME: 25–30 MINUTES
SERVES 4

700 g (1 lb 9 oz) orange sweet potato, peeled and cut into 1.5 cm (5/8 inch) dice

60 ml (2 fl oz/1/4 cup) olive oil

200 g (7 oz/1 cup) quinoa (see tip), briefly rinsed

1/2 teaspoon ground turmeric

1 teaspoon ground cumin

250 g (9 oz) broccoli, trimmed and cut into small florets

1 bunch asparagus, trimmed and cut into 2 cm (3/4 inch) lengths

50 g (1 1/2 oz/1/3 cup) sweetened dried cranberries

1/3 cup coarsely chopped mixed herbs (such as chives, parsley, mint)

2 tablespoons fresh orange juice

1 tablespoon white balsamic vinegar

* *The sweet potato and quinoa give this salad a low GI rating. If you have any leftovers, take them to work for a lunch that will provide slow-release energy throughout the day.*

1 Preheat the oven to 200°C (400°F/Gas 6). Place the sweet potato on a large baking tray lined with non-stick baking paper. Drizzle with 1 tablespoon of the olive oil and season with sea salt and pepper. Roast for 25–30 minutes or until golden and tender.

2 Meanwhile, put the quinoa, turmeric, cumin and 500 ml (17 fl oz/2 cups) water in a medium saucepan and bring to the boil. Reduce the heat to low, cover and simmer for 12 minutes or until the water is absorbed. Remove from the heat and set aside to cool slightly.

3 Blanch the broccoli and asparagus in a saucepan of boiling water until tender-crisp, then drain and refresh under cold running water.

4 Put the cooked quinoa, sweet potato, blanched vegetables, cranberries and herbs in a large bowl. Whisk together the remaining 2 tablespoons of olive oil, the orange juice and balsamic. Add to the salad and toss to combine, then season with sea salt and freshly ground black pepper, to taste.

TIP We used equal quantities of red and white quinoa in this recipe. There is no need to cook them separately.

PARMESAN & HERB CHICKEN SCHNITZELS WITH TOMATO SALSA

(PAGE 26)

Parmesan & herb chicken schnitzels with tomato salsa

PREPARATION TIME: 20 MINUTES
COOKING TIME: 15 MINUTES
SERVES 4

110 g (3¾ oz/1 cup) quinoa flakes
35 g (1¼ oz/⅓ cup) finely grated parmesan cheese
2 tablespoons finely snipped chives
2 tablespoons finely chopped flat-leaf (Italian) parsley
1 teaspoon finely grated lemon zest
35 g (1¼ oz/¼ cup) plain (all-purpose) flour (see tip)
2 eggs, lightly whisked
60 ml (2 fl oz/¼ cup) milk
600 g (1 lb 5 oz) chicken tenderloins
2 tablespoons olive oil, approximately
Salad leaves or steamed greens, to serve

TOMATO SALSA
3 ripe roma (plum) tomatoes, diced
½ teaspoon dried red chilli flakes
2 teaspoons red wine vinegar
1½ tablespoons olive oil

1 Combine the quinoa flakes, parmesan, herbs and lemon zest in a shallow bowl. Put the flour in a second shallow bowl and season with salt and pepper. Put the combined eggs and milk in a third bowl.

2 Dust a tenderloin in the seasoned flour, then dip in the egg mixture and finally press into the quinoa mixture until it is evenly coated. Place on a tray and repeat to coat the remaining tenderloins.

3 To make the tomato salsa, put all the ingredients in a medium bowl and stir to combine. Season with a pinch of sugar, sea salt and freshly ground black pepper, to taste. Set aside.

4 Heat the oil in a large non-stick frying pan over medium–high heat. Cook the crumbed chicken, in batches, for 3 minutes each side or until golden and cooked through, adding a little more oil to the pan as necessary.

5 Serve the chicken schnitzels with the tomato salsa and salad leaves or steamed greens.

TIP If you would like these delicious schnitzels to be gluten free, simply replace the plain flour with rice flour.

* *Quinoa flakes are fantastic for crumbing meat, chicken, fish, fishcakes or patties. They stick easily, become crisp and golden once pan-fried, and have more flavour than breadcrumbs.*

Red lentil & quinoa dhal

PREPARATION TIME: 20 MINUTES
COOKING TIME: 35–40 MINUTES
SERVES 4

1 tablespoon sunflower oil
1 red onion, finely chopped
2 teaspoons finely grated ginger
1 long red chilli, seeded and finely chopped
2 garlic cloves, crushed
1 teaspoon ground coriander
2 teaspoons ground cumin
12 fresh curry leaves (see tips), plus extra, to garnish
100 g (3½ oz/½ cup) split red lentils, briefly rinsed
100 g (3½ oz/½ cup) quinoa (see tips), briefly rinsed
1 tablespoon lime juice
Caster (superfine) sugar, to taste
Plain yoghurt, to garnish
Poppadoms (see tips), to serve

* *Adding quinoa to dhal makes this dish ideal for vegetarians or vegans, as lentils and quinoa are complementary proteins.*

1 Heat the oil in a large saucepan over medium heat. Add the onion and cook, stirring occasionally, for 5 minutes or until soft. Add the ginger, chilli, garlic, spices and curry leaves and cook for 1–2 minutes or until fragrant.

2 Add the lentils, quinoa and 750 ml (26 fl oz/3 cups) water and bring to the boil. Reduce the heat to low and simmer, stirring occasionally, for 25–30 minutes or until the lentils and quinoa are soft but still have some texture, and the mixture is thick. Stir through the lime juice and season with sugar, sea salt and freshly ground black pepper, to taste.

3 Garnish with the yoghurt and extra curry leaves, and serve with the poppadoms.

TIPS Fresh curry leaves can be found in the herb section at larger supermarkets, and good fruit and vegetable stores.

We used red quinoa in this recipe.

If you do not require this meal to be gluten free, you can serve it with naan bread instead of the poppadoms.

If you like your dhal really spicy, try adding ½ teaspoon dried red chilli flakes with the spices in step 1.

This dhal makes a fantastic dip to serve at parties, and can be eaten warm or at room temperature.

Pancetta, spinach & leek frittata

PREPARATION TIME: 20 MINUTES
COOKING TIME: 50–55 MINUTES
SERVES 6

100 g (3½ oz/½ cup) quinoa (see tips), briefly rinsed
1 tablespoon olive oil
100 g (3½ oz) pancetta (see tips), diced
1 large leek, white part only, thinly sliced
1 garlic clove, crushed
200 g (7 oz) baby spinach leaves
8 eggs
185 ml (6 fl oz/¾ cup) milk
35 g (1¼ oz/⅓ cup) finely grated parmesan cheese
75 g (2¾ oz) goat's cheese, crumbled
1 tablespoon snipped chives
Rocket (arugula) leaves, to serve

* *Quinoa is a great source of slow-release carbohydrate. Just add some salad leaves to make this frittata a balanced meal, containing carbohydrate, protein and vegetables.*

1 Preheat the oven to 180°C (350°F/Gas 4). Lightly grease a 20 x 30 cm (8 x 12 inch) rectangular baking dish and line the base with non-stick baking paper. Put the quinoa in a small saucepan with 250 ml (9 fl oz/1 cup) water and bring to the boil. Reduce the heat to low, cover and simmer for 12 minutes or until the water is absorbed. Remove from the heat and set aside to cool slightly.

2 Heat the oil in a medium frying pan over medium heat. Cook the pancetta and leek, stirring occasionally, for 5 minutes or until golden. Add the garlic and cook, stirring, for 1 minute. Remove from the heat and set aside.

3 Put the spinach in a large heatproof bowl and pour over enough boiling water to cover. Blanch for 30 seconds, then drain and refresh under cold running water. Drain well, squeezing out any excess water. Coarsely chop.

4 Whisk the eggs and milk together in a large bowl. Stir through the quinoa, leek mixture, spinach and parmesan. Pour into the prepared dish, dot the goat's cheese over the top and sprinkle with the chives. Bake for 30–35 minutes or until the frittata is set, puffed and golden.

5 Serve the frittata warm or cold, cut into squares and accompanied by a rocket salad.

TIPS We used white quinoa in this recipe.
Some pancetta may contain gluten, so always check the label.
This frittata is ideal for lunch boxes and picnic baskets.
To make a richer and creamier version of this frittata, replace the milk with thin (pouring/whipping) cream.

Quinoa, feta, roasted capsicum & corn muffins

PREPARATION TIME: 20 MINUTES
COOKING TIME: 20 MINUTES
MAKES 12

225 g (8 oz/1½ cups) self-raising flour
110 g (3¾ oz/1 cup) quinoa flakes
200 g (7 oz/1 cup) corn kernels (fresh, tinned or frozen)
120 g (4¼ oz) drained roasted capsicum (pepper) in oil, diced
75 g (2¾ oz) feta cheese, crumbled
2 tablespoons snipped chives
250 ml (9 fl oz/1 cup) milk
1 egg
80 ml (2½ fl oz/⅓ cup) sunflower oil

* *I love adding quinoa flakes to baked goods such as these muffins. They taste great and also add protein, calcium, iron and B vitamins, making them the perfect addition to lunch boxes.*

1 Preheat the oven to 180°C (350°F/Gas 4). Line a 12-hole 80 ml (2½ fl oz/⅓ cup) muffin tin with paper cases.

2 Sift the flour into a large bowl and stir through the quinoa flakes, corn, capsicum, feta and chives. Combine the milk, egg and oil in a large jug, add to the dry ingredients and stir until well combined (do not over-mix).

3 Divide the mixture among the lined muffin holes and bake for 20 minutes or until golden and cooked through. Set aside for 5 minutes to cool, then transfer to a wire rack. Serve warm or at room temperature.

TIP These muffins can be frozen for up to 1 month, wrapped well in plastic wrap. Thaw at room temperature.

Vegetables stuffed with quinoa, lamb, pine nuts & currants

PREPARATION TIME: 20 MINUTES
COOKING TIME: 50–55 MINUTES
SERVES 4

100 g (3½ oz/½ cup) quinoa, briefly rinsed

4 large (about 125 g/4½ oz each) vine-ripened tomatoes

4 zucchini (courgettes) (about 200 g/7 oz each)

250 g (9 oz) minced (ground) lamb

1 teaspoon ground allspice

4 spring onions (scallions), trimmed and thinly sliced

2 tablespoons currants

2 tablespoons coarsely chopped mint

2 tablespoons coarsely chopped flat-leaf (Italian) parsley

40 g (1½ oz/¼ cup) pine nuts, lightly toasted

75 g (2¾ oz) creamy feta cheese, crumbled

Salad leaves, to serve

* *Stuffed vegetables traditionally contain breadcrumbs, couscous or cracked wheat. Quinoa is a great gluten-free alternative and it works beautifully with the Middle Eastern flavours in this recipe.*

1 Preheat the oven to 180°C (350°F/Gas 4). Line a large baking tray with non-stick baking paper. Put the quinoa in a small saucepan with 250 ml (9 fl oz/1 cup) water and bring to the boil. Reduce the heat to low, cover and simmer for 12 minutes or until the water is absorbed. Remove from the heat and set aside to cool completely.

2 Meanwhile, cut off the top of each tomato. Scoop out the flesh and pulp and reserve, leaving a 1 cm (½ inch) thick shell. Halve each zucchini lengthways. Scoop out the flesh and discard, leaving a 1 cm (½ inch) thick shell.

3 Coarsely chop the reserved tomato flesh and put in a large bowl with the reserved tomato pulp, cooked quinoa, minced lamb, allspice, spring onions, currants, herbs, pine nuts and feta. Stir until all the ingredients are well combined, then season with sea salt and freshly ground black pepper.

4 Stuff each tomato and zucchini shell with the quinoa mixture. Place the stuffed zucchini on the lined tray and bake for 15 minutes. Add the stuffed tomatoes to the tray of zucchini and bake for a further 20–25 minutes or until the zucchini and tomatoes are tender and the stuffing is golden. Serve with salad leaves.

TIPS We used white quinoa in this recipe.

For a vegetarian version, omit the lamb, increase the quinoa to 200 g (7 oz/1 cup) and increase the water to 500 ml (17 fl oz/2 cups).

Chunky chocolate, quinoa & pecan cookies

PREPARATION TIME: 15 MINUTES
COOKING TIME: 12–15 MINUTES
MAKES ABOUT 18

130 g (4½ oz) unsalted butter, softened

125 g (4½ oz/⅔ cup, lightly packed) light brown sugar

1 egg, lightly whisked

1 teaspoon natural vanilla extract

110 g (3¾ oz/¾ cup) plain (all-purpose) flour

1 teaspoon baking powder

110 g (3¾ oz/1 cup) quinoa flakes

150 g (5½ oz) milk chocolate, chopped

65 g (2¼ oz/⅔ cup) pecan nuts, chopped

** The combination of chocolate, pecans and brown sugar with the nutty taste of quinoa flakes makes these cookies absolutely delicious.*

1 Preheat the oven to 180°C (350°F/Gas 4) and line 2 baking trays with non-stick baking paper.

2 Use an electric mixer to cream the butter and sugar until pale and creamy. Add the egg and vanilla and beat until well combined. Sift the flour and baking powder together, add to the butter mixture and stir until combined. Add the quinoa flakes, chocolate and pecan nuts and stir until just combined.

3 Roll rounded tablespoons of the mixture into balls and place on the lined trays, 4 cm (1½ inches) apart. Flatten the balls with a fork dipped in flour, then bake for 12–15 minutes or until pale golden.

4 Remove from the oven and cool on the trays for 5 minutes before transferring to a wire rack to cool completely.

TIPS For a gluten-free version, replace the plain flour with 55 g (2 oz/⅓ cup) rice flour, and 55 g (2 oz) gluten-free plain (all-purpose) flour.

For a nut-free version, replace the pecans with chopped dates or sweetened dried cranberries.

Store these cookies in an airtight container for up to 5 days.

Carrot quinoa cake with cream cheese frosting

PREPARATION TIME: 20 MINUTES (+ COOLING)
COOKING TIME: 50 MINUTES
SERVES 12

190 g (6¾ oz/1½ cups) gluten-free self-raising flour (such as Orgran brand)

1 teaspoon ground cinnamon

½ teaspoon ground nutmeg

180 g (6¼ oz/1 cup, lightly packed) light brown sugar

50 g (1¾ oz/½ cup) almond meal

185 g (6½ oz/1 cup) cooked quinoa (see page 15)

235 g (8½ oz/1½ cups, firmly packed) grated carrot

85 g (3 oz/½ cup) seedless raisins, coarsely chopped

185 ml (6 fl oz/¾ cup) vegetable oil

3 eggs

FROSTING

250 g (9 oz) cream cheese, softened

1 teaspoon finely grated orange zest

40 g (1½ oz/⅓ cup) pure icing (confectioners') sugar, sifted

2 tablespoons freshly squeezed orange juice, strained

* *The cooked quinoa gives this cake a delicious moist texture.*

1 Preheat the oven to 180°C (350°F/Gas 4). Grease a square 20 cm (8 inch) cake tin and line with non-stick baking paper.

2 Sift the flour, cinnamon and nutmeg together into a large mixing bowl. Add the sugar, almond meal, quinoa, carrot and raisins and stir to combine. Whisk the oil and eggs together, add to the quinoa mixture and stir until well combined.

3 Spoon the batter into the prepared tin and bake for 50 minutes or until the cake is firm to touch and a skewer inserted into the centre comes out with only a few moist crumbs clinging to it. (You may need to cover the cake with foil if it is browning too quickly.) Transfer the tin to a wire rack and set aside to cool for 15 minutes. Remove the cake from the tin and set aside on the wire rack to cool completely.

4 To make the frosting, use an electric mixer to beat the cream cheese and orange zest until smooth. Add the icing sugar and orange juice and beat until well combined. Spread evenly over the top and sides of the cooled carrot cake and serve.

TIP If you don't require this cake to be gluten free, use 225 g (7¾ oz/1½ cups) wheat self-raising flour instead of the gluten-free flour.

This cake will keep, stored in an airtight container in a cool place, for up to 3 days.

Amaranth

NUTRITIONAL INFORMATION
(per 245 g/9 oz/1 cup cooked amaranth)

Energy: 1051 kilojoules, 251 calories **Protein:** 9.3 g **Total fat:** 3.9 g
Saturated fat: 0 g **Carbohydrate:** 46 g **Sugars:** 0 g **Dietary fibre:** 5.2 g
Cholesterol: 0 mg **Vitamin B_6:** 0.3 mg **Iron:** 5.2 mg **Magnesium:** 160 mg
Manganese: 2.1 mg **Phosphorus:** 364 mg **Sodium:** 14.8 mg

Amaranth is a pseudo-grain, like quinoa, grown by the ancient Aztecs over 5000 years ago. Its name is derived from the Greek *amarantos,* which means 'unfading'. Amaranth was a staple crop not only to the Aztecs, but also the Incas and Mayans, until its cultivation was banned by the invading Spanish conquistadors as it was often used in local religious rituals.

Amaranth plants are a member of the goosefoot family, and therefore related to spinach. They produce tiny edible seeds, with up to 500,000 per plant, and their leaves are also highly nutritious and edible.

When amaranth experienced a revival in the 1970s, wild varieties found in Mexico were grown for commercial use. Today, amaranth is again growing in popularity. The recent attention it is receiving is mainly due to its impressive nutritional qualities. It is gluten and wheat free, so it is suitable for those with coeliac disease or wheat intolerances. It is also a good source of vegetable protein (13–14%) due to its amino acid balance, particularly the presence of lysine, an amino acid that is missing or negligible in most other grains.

Amaranth grain has a glycaemic index of just 21, so it is considered a low-GI food. This makes it ideal for diabetics or those following a low-GI diet. It is also rich in phytosterols, which are linked to reducing cholesterol.

Amaranth is packed with vitamins and minerals, in particular calcium, iron, magnesium, manganese, phosphorus, vitamin E, and B vitamins. Amaranth also contains good fats, with an oil content of 6–9% which is higher than most other grains. Amaranth oil is high in unsaturated fatty acids, especially linoleic acid.

Amaranth is available as a whole grain, puffed and as flour. Amaranth grains are tiny, similar in size and colour to brown mustard seeds. Like quinoa it has a central seed and an outer germ that uncurls during cooking.

Amaranth grain cooks relatively quickly, in about 20 minutes, however it is a little trickier and less forgiving than quinoa, which explains why it has not gained as much popularity. Once cooked, it releases a starch and develops a slightly sticky texture on standing, so it's best combined with another grain, such as brown rice, millet or quinoa, if you want to use it in salads. As a general rule, I'd suggest using one part amaranth to three parts another grain. The natural stickiness of amaranth does work in its favour when making porridge, however.

Amaranth flour can be used in gluten-free baked goods that do not require leavening such as flat breads, brownies and cookies. It can also be substituted for approximately one-quarter of the flour in leavened baked products such as bread. It has quite a distinct herbaceous flavour that can be a little overpowering on its own.

Puffed amaranth is my favourite form of amaranth, and it has the mildest flavour of the three. It gives a delicious crisp texture when added to muesli and breakfast cereals, biscuits, muesli bars and other baked goods. You can make puffed amaranth yourself, but it tends to burn easily, so it is much easier to buy it already made. A common sweet still made in Mexico today is alegria, made from puffed amaranth, honey and chocolate.

To cook amaranth

Toasting amaranth grain before cooking tends to soften the flavour slightly and is generally preferred. Place 210 g (7½ oz/1 cup) amaranth in a frying pan over medium heat. Cook, stirring, for 3 minutes or until just fragrant. Transfer to a saucepan, add 625 ml (21½ fl oz/2½ cups) water and bring to the boil. Reduce the heat to low, cover and simmer for 20 minutes or until the grain is *al dente* and the mixture has thickened.

Andean grain salad with salmon, red grapes & watercress

PREPARATION TIME: 20 MINUTES
(+ COOLING)
COOKING TIME: 20 MINUTES
SERVES 4

200 g (7 oz/1 cup) Andean grain mix (see tips), briefly rinsed
2 x 200 g (7 oz) skinless salmon fillets
2 tablespoons olive oil, plus 2 teaspoons, extra
250 g (9 oz) sugar snap peas, trimmed
200 g (7 oz) seedless red grapes, halved
60 g (2¼ oz/2 cups) picked watercress sprigs (see tips), washed
100 g (3½ oz) creamy feta cheese, crumbled
¼ cup snipped chives
1½ tablespoons lemon juice
1 teaspoon finely grated lemon zest
½ teaspoon caster (superfine) sugar

** Amaranth grain is rich in the minerals manganese, iron, magnesium and phosphorus. It is also packed with dietary fibre and is a great vegetable source of protein. It becomes slightly sticky when cooked so it is best combined with quinoa when used as a salad grain.*

1 Put the grain mix in a medium saucepan with 500 ml (17 fl oz/ 2 cups) water and bring to the boil. Reduce the heat to low, cover and simmer for 12 minutes or until the water is absorbed. Remove from the heat and set aside to cool completely.

2 Brush the salmon fillets with the extra olive oil. Heat a medium non-stick frying pan over high heat and cook the salmon for 2–3 minutes each side (for medium) or until cooked to your liking. Remove and set aside to cool slightly, then break up the salmon into large flakes.

3 Blanch the sugar snaps in a saucepan of boiling water until they are tender-crisp. Drain and refresh under cold running water.

4 Put the cooked grain mix, salmon, sugar snaps, grapes, watercress, feta and chives in a large bowl. Whisk the olive oil, lemon juice, zest and sugar together. Add to the salad and gently toss to combine, then season with sea salt and freshly ground black pepper to taste.

TIPS Andean grain mix is also sometimes called 'supergrain mix' and is a combination of amaranth and red, white and black quinoa. To make your own, simply combine equal quantities of these grains.

You can replace the watercress leaves with baby rocket (arugula) or spinach leaves if you like.

Amaranth & raisin cookies

PREPARATION TIME: 15 MINUTES
COOKING TIME: 15 MINUTES
MAKES ABOUT 24

80 g (2¾ oz) unsalted butter
50 g (1¾ oz) crunchy peanut butter
2 tablespoons honey
150 g (5½ oz/1 cup) plain (all-purpose) flour (see tips)
½ teaspoon ground ginger
100 g (3½ oz/2 cups) puffed amaranth
75 g (2¾ oz/⅓ cup) caster (superfine) sugar
75 g (2¾ oz/⅓ cup, firmly packed) light brown sugar
85 g (3 oz/½ cup) raisins
1 egg, lightly whisked

** Puffed amaranth adds a delicious crisp texture to baked goods such as cookies and muesli bars.*

1 Preheat the oven to 180°C (350°F/Gas 4) and line 2 large baking trays with non-stick baking paper. Put the butter, peanut butter and honey in a small saucepan over medium heat and stir occasionally until the butter has melted. Set aside to cool slightly.

2 Sift the flour and ginger into a large mixing bowl. Add the amaranth, sugars, raisins, cooled butter mixture and egg and stir until well combined.

3 Using slightly wet hands, roll rounded tablespoons of the mixture into balls and place on the lined trays, 4 cm (1½ inches) apart. You need to roll each ball in your hand until it becomes sticky, about 30 seconds. Use a lightly floured fork to flatten the balls slightly, then bake for 10–12 minutes or until golden.

4 Remove from the oven and cool on the trays for 5 minutes before transferring to a wire rack to cool completely.

TIPS If you would like to make these cookies gluten-free, substitute the plain flour with amaranth flour, or use 75 g (2¾ oz) rice flour and 75 g (2¾ oz) gluten-free plain (all-purpose) flour.

These cookies will keep in an airtight container for up to 5 days.

AMARANTH & MILLET PORRIDGE WITH PISTACHIO SEED SPRINKLE (PAGE 46)

gluten-free

Amaranth & millet porridge with pistachio seed sprinkle

PREPARATION TIME: 10 MINUTES
COOKING TIME: 30–35 MINUTES
SERVES 4

160 g (5½ oz/¾ cup) hulled millet
55 g (2 oz/¼ cup) amaranth
½ teaspoon ground cinnamon
250 ml (9 fl oz/1 cup) milk
40 g (1½ oz) dried apricots, chopped
Greek-style yoghurt and honey, to serve

PISTACHIO SEED SPRINKLE
45 g (1¾ oz/⅓ cup) pistachio kernels, chopped
40 g (1½ oz/¼ cup) pepitas (pumpkin seeds)
40 g (1½ oz/¼ cup) sunflower seeds
2 tablespoons sesame seeds
1 tablespoon light brown sugar

* *Amaranth is well suited for making porridge because it becomes slightly sticky when cooked with liquid. Amaranth and millet have similar cooking times so are an ideal combination for a gluten-free breakfast.*

1 Preheat the oven to 180°C (350°F/Gas 4) and line a baking tray with non-stick baking paper. To make the pistachio seed sprinkle, put all the ingredients in a bowl and stir to combine. Spread the mixture over the lined tray and bake for 5–10 minutes or until golden. Set aside to cool.

2 Meanwhile, put the millet, amaranth and cinnamon in a large saucepan with 625 ml (21½ fl oz/2½ cups) water and bring to the boil. Reduce the heat to low, cover and simmer, stirring occasionally, for 20 minutes or until the water is absorbed. Add the milk and apricots, cover and simmer, stirring occasionally, for 5–10 minutes or until the porridge is thick and creamy and the grains are tender.

3 Serve the porridge topped with a spoonful of yoghurt, a drizzle of honey and some pistachio seed sprinkle.

TIP The pistachio seed sprinkle will keep in an airtight container for up to 2 weeks. It is also delicious sprinkled over fruit and yoghurt.

Amaranth, cashew & apricot balls

PREPARATION TIME: 15 MINUTES
COOKING TIME: 2 MINUTES
MAKES ABOUT 20

95 g (3¼ oz/1 cup) rolled (porridge) oats
55 g (2 oz/⅓ cup) sweetened dried cranberries
45 g (1¾ oz/⅓ cup) chopped dried apricots
40 g (1½ oz) dark chocolate, chopped
50 g (1¾ oz/1 cup) puffed amaranth
65 g (2¼ oz/¼ cup) cashew spread
115 g (4 oz/⅓ cup) honey
45 g (1¾ oz/½ cup) desiccated coconut

* *Puffed amaranth can be used in recipes where you would typically use plain puffed rice cereal, such as muesli bars, biscuits and slices. Amaranth is rich in protein and dietary fibre, and is easily digested as it is gluten free. Here it's combined with rolled oats, nuts and dried fruit, making little snacks that are packed with nutrients.*

1 Line a large baking tray with non-stick baking paper. Put the oats, cranberries, apricots and chocolate in a food processor and process until finely chopped. Transfer to a large mixing bowl and mix in the puffed amaranth.

2 Put the cashew spread and honey in a small saucepan over low heat. Warm gently and stir until combined. Add the honey mixture to the dry ingredients and mix until well combined.

3 Place the coconut on a plate. Using slightly wet hands (to prevent sticking) roll tablespoons of the mixture into balls, then press in the coconut to lightly coat. Cover and refrigerate until firm.

TIPS These are perfect for after-school snacks or lunch boxes.
For a nut-free version, replace the cashew spread with tahini or nut-free butter.

Gluten-free muesli with dried apple, pepitas & almonds

PREPARATION TIME: 15 MINUTES
COOKING TIME: NIL
MAKES ABOUT 6½ CUPS

75 g (2¾ oz/1½ cups) puffed amaranth
22 g (¾ oz/1½ cups) puffed millet
80 g (2¾ oz/1 cup) rice bran flakes
100 g (3½ oz/½ cup) roasted buckwheat
50 g (1¾ oz) dried apple, chopped
85 g (3 oz) sweetened dried cranberries
100 g (3½ oz) almonds, coarsely chopped
40 g (1½ oz/½ cup) pepitas
2 tablespoons linseeds
40 g (1½ oz/¼ cup) sunflower seeds

** This gluten-free muesli runs rings around the commercial varieties, which can often be bland. The puffed amaranth and millet add lots of flavour, while the roasted buckwheat gives body, texture and a nutty taste.*

This muesli is packed with protein, complex carbohydrates and dietary fibre, plus the nuts and seeds provide good oils, just what you need to start your day.

1 Put all the ingredients in a large bowl and mix to combine. Transfer to an airtight container and store until needed.

TIPS For a wheat-free version, add some rolled (porridge) oats.

This muesli will keep for about 2 weeks in an airtight container in a cool dark place.

Little passionfruit & coconut cakes

PREPARATION TIME: 15 MINUTES
COOKING TIME: 20 MINUTES
MAKES 12

240 g (8¾ oz) pure icing (confectioners') sugar, sifted, plus extra, to serve

65 g (2¼ oz/½ cup) amaranth flour

½ teaspoon baking powder (see tips)

100 g (3½ oz/1 cup) almond meal

50 g (1¾ oz) desiccated coconut

6 egg whites, lightly whisked

120 g (4¼ oz) unsalted butter, melted and cooled

1 teaspoon finely grated lime zest

70 g (2½ oz/¼ cup) passionfruit pulp (see tips)

** Amaranth flour has quite a herbaceous flavour, so it works well with the nuttiness of almond meal. Amaranth flour is gluten free, so it is ideal to use in gluten-free baking.*

1 Preheat the oven to 180°C (350°F/Gas 4). Line a 12-hole 80 ml (2½ fl oz/⅓ cup) muffin tin with paper cases.

2 Sift the icing sugar, amaranth flour and baking powder together into a large mixing bowl. Stir through the almond meal and coconut.

3 Add the egg whites, butter, lime zest and passionfruit pulp and stir to combine. Divide the mixture among the lined muffin holes.

4 Bake for 20 minutes or until golden and cooked through (a skewer inserted into the centre of a cake should come out with a few moist crumbs clinging to it). Remove from the oven and set aside for 5 minutes, then transfer to a wire rack to cool. Serve dusted with extra icing sugar.

TIPS If you want these cakes to be gluten free, make sure you use gluten-free baking powder.

You will need about 3 passionfruit to get this amount of pulp.

The cakes will keep, in an airtight container, for up to 3 days.

Buckwheat

NUTRITIONAL INFORMATION
(per 170 g/5¾ oz/1 cup cooked buckwheat)

Energy: 649 kilojoules, 155 calories **Protein:** 5.7 g **Total fat:** 1 g
Saturated fat: 0.2 g **Carbohydrate:** 33.5 g **Sugars:** 1.5 g **Dietary fibre:** 4.5 g
Cholesterol: 0 mg **Iron:** 1.3 mg **Magnesium:** 85.7 mg **Manganese:** 0.7 mg
Phosphorus: 118 mg **Sodium:** 6.7 mg

Buckwheat is not a type of wheat, but is the triangular seed of a plant related to rhubarb and sorrel. Although technically a pseudo-grain, it is widely classified and utilised as a grain, and is a great substitute for those sensitive to wheat or gluten. Buckwheat is also quick to prepare, a high-quality vegetable protein and is rich in various nutrients and phytochemicals, making it a wonderful addition to any diet.

Buckwheat is an ancient crop, first cultivated in South-East Asia around 6000BC, from where it spread to Central Asia, Europe and the Middle East. It grows better in cooler climates and is a short-season crop. In the mountainous regions of northern China and Tibet, where wheat cannot be grown, buckwheat has long been used to make buckwheat noodles. Today, the top producers of buckwheat are Russia and China. One of the best known uses of buckwheat would have to be in the small buckwheat pancakes, called blini, originally from Russia. Buckwheat porridge made from the grains (also known as groats) is a peasant dish in eastern Europe.

While buckwheat (soba) noodles and pancakes may be familiar to many, buckwheat grain tends to be under-utilised in Western countries. This is mainly due to a lack of knowledge on how to use and prepare it properly. However, a growing awareness of its unique nutritional qualities is gradually increasing its popularity, especially with those following a wheat-free, gluten-free or vegetarian diet.

Buckwheat is a quality source of a complete vegetable protein (13–15%), containing all essential amino acids, so it is an ideal food for vegetarians and vegans. It is an excellent source of manganese and magnesium, and a good source of niacin, folate, iron, zinc, copper, selenium and phosphorus.

Buckwheat is also rich in phytochemicals. Of particular interest is the antioxidant rutin, which is believed to help lower cholesterol and reduce blood pressure. It is a great source of polyunsaturated fatty acids, such as linoleic acid, as well as being high in soluble fibre, aiding gastrointestinal health.

Buckwheat can be purchased in its hulled grain form (known as groats), as a flour or as puffed buckwheat. It can also be purchased in its processed form, as pasta or noodles. It has a distinct earthy flavour that can polarise people, but it definitely grows on you the more you eat it.

Buckwheat kernels are triangular in shape and have a hard inedible hull that must be removed. The groats can be purchased either raw or toasted. Toasted buckwheat can have quite a strong flavour that often disguises stale buckwheat. I find buying the raw version and toasting it myself gives a milder and far superior result. Buckwheat kernels can be used in salads, porridge or as an alternative to rice in risotto and pilafs.

Buckwheat flour is made from grinding the kernels and has a strong earthy flavour. It is used to make noodles, pancakes and bread, and is often combined with some wheat flour to tone down the taste. Puffed buckwheat is a delicious addition to home-made muesli, or baked goods such as muesli bars.

To cook buckwheat

To toast buckwheat, heat a large non-stick frying pan over medium–high heat. Add 200 g (7 oz/1 cup) raw buckwheat kernels and cook, stirring, for 3–4 minutes or until just fragrant.

To cook, put the toasted buckwheat and 500 ml (17 fl oz/2 cups) water in a saucepan and bring to the boil. Cover, reduce the heat to low and simmer for 10–12 minutes or until just tender (be careful not to overcook). Rinse under cold running water; drain well.

gluten-free

Roasted beetroot, buckwheat & goat's cheese salad

PREPARATION TIME: 20 MINUTES (+ COOLING)
COOKING TIME: 45 MINUTES
SERVES 4

10–12 (about 600 g/1 lb 5 oz in total) baby beetroot (beets), trimmed
200 g (7 oz/1 cup) raw buckwheat
75 g (2¾ oz/½ cup) hazelnuts, lightly toasted and skinned
100 g (3½ oz) baby rocket (arugula) leaves
2 tablespoons snipped chives
100 g (3½ oz) soft goat's cheese (see tips), crumbled

DRESSING

2 tablespoons olive oil
2 tablespoons fresh orange juice
1 tablespoon balsamic vinegar

** The earthy flavour of the buckwheat is delicious with roasted beetroot and goat's cheese. Buckwheat is an ideal grain for vegetarian dishes such as this, as it is rich in protein, B vitamins, calcium and phosphorus. The cheese and hazelnuts also add extra protein.*

1 Preheat the oven to 200°C (400°F/Gas 6). Put the beetroot in a large roasting pan, cover the pan with foil and roast for 40–45 minutes or until they are tender when pierced with a skewer. Set aside to cool slightly, then peel (see tips) and cut into wedges.

2 Meanwhile, heat a large non-stick frying pan over medium–high heat. Add the buckwheat and cook, stirring, for 3–4 minutes or until fragrant. Put the toasted buckwheat and 500 ml (17 fl oz/2 cups) water in a saucepan and bring to the boil. Cover, reduce the heat to low and simmer for 10–12 minutes or until just tender (be careful not to overcook). Rinse under cold running water, then drain well.

3 Coarsely chop the hazelnuts and place in a large bowl. Add the beetroot, cooked buckwheat, rocket and chives and stir to combine.

4 To make the dressing, whisk the olive oil, orange juice and balsamic in a small bowl until combined. Add to the salad and toss well. Sprinkle with the goat's cheese and season with sea salt and freshly ground black pepper, to taste.

TIPS Wear gloves while peeling the beetroot to prevent your hands becoming stained by the juices.

You could substitute a creamy feta cheese or marinated feta for the goat's cheese if you like.

Soba noodles with beef & miso broth

PREPARATION TIME: 20 MINUTES
COOKING TIME: 25 MINUTES
SERVES 4

750 ml (26 fl oz/3 cups) beef stock
2 teaspoons finely grated ginger
2 tablespoons soy sauce
1 tablespoon mirin
180 g (6¼ oz) dried soba noodles
1 tablespoon sunflower oil
1 large white onion, thinly sliced
2 tablespoons white or yellow (brown) miso paste
1 bunch baby bok choy (pak choy), trimmed
300 g (10½ oz) beef eye fillet, thinly sliced (see tip)
Thinly sliced spring onion (scallion), to garnish

Soba noodles are a great way to incorporate buckwheat into your diet. Most soba noodles contain wheat and buckwheat flour, so are not suitable for a gluten-free diet. You can buy gluten-free soba noodles made from 100% buckwheat from health food stores, but they are much stronger in flavour and do tend to stick a little, so make sure you rinse them well once cooked.

1 Put the stock, 750 ml (26 fl oz/3 cups) water, the ginger, soy sauce and mirin in a large saucepan and bring to the boil. Reduce the heat to low and simmer for 5 minutes. Remove from the heat and set aside.

2 Cook the noodles in a large saucepan of salted boiling water, following the packet instructions, until *al dente*. Drain well, then divide among 4 serving bowls.

3 Clean the saucepan and return to medium heat. Add the oil and cook the onion, stirring, for 5 minutes or until softened. Add the stock mixture and bring to a simmer, then whisk in the miso paste until smooth. Add the bok choy and simmer for 1 minute, then remove from the heat. Add the beef and stir to combine (the heat of the soup will cook the beef).

4 Ladle the hot soup over the noodles and serve garnished with the spring onion.

TIP Partially freezing the beef makes it easier to slice it paper-thin.

Buckwheat pasta with pancetta, broccoli & chilli

PREPARATION TIME: 20 MINUTES
COOKING TIME: 20 MINUTES
SERVES 4

600 g (1 lb 5 oz) broccoli, trimmed and cut into florets

25 g (1 oz/1/4 cup) finely grated parmesan cheese, plus extra, to serve

300 g (10 1/2 oz) buckwheat pasta (we used spirals)

1 tablespoon olive oil

100 g (3 1/2 oz) pancetta (see tips), diced

2 garlic cloves, crushed

1 teaspoon finely grated lemon zest

1/2 teaspoon dried red chilli flakes

80 ml (2 1/2 fl oz/1/3 cup) thin (pouring/whipping) cream

** Buckwheat pasta is gluten free, so it is a great alternative to wheat pasta for those on a wheat-free or gluten-free diet. It is available from health food stores and some supermarkets. It has quite an earthy, nutty taste that is delicious with this simple broccoli sauce. Buckwheat pasta tends to cook a little faster than regular pasta — be careful not to overcook it as it doesn't hold its shape as well as wheat pasta.*

1 Cook the broccoli in a large saucepan of boiling water until bright green and tender-crisp. Drain, reserving 80 ml (2 1/2 fl oz/1/3 cup) of the cooking liquid. Refresh the broccoli under cold running water, then drain well.

2 Put half the broccoli in a food processor (reserve the remaining broccoli) and add 2 tablespoons of the reserved cooking liquid and the parmesan. Process until smooth, adding a little more of the cooking liquid if necessary.

3 Cook the pasta in salted boiling water according to the packet instructions, until *al dente*, then drain well.

4 Meanwhile, heat the oil in a large non-stick frying pan and cook the pancetta, stirring occasionally, for 3–4 minutes or until golden. Add the garlic, zest and chilli and cook, stirring, for 1 minute more. Stir in the reserved broccoli florets, the processed broccoli mixture and the cream and simmer for 2 minutes.

5 Add the pasta and toss together for 1–2 minutes or until the pasta is coated in the sauce and heated through. Season with sea salt and freshly ground black pepper and serve garnished with extra parmesan.

TIPS Some pancetta may contain gluten, so always check the label.
You can use spelt, kamut or wholemeal (whole-wheat) pasta instead of buckwheat pasta, but the dish will not be gluten free.

BUCKWHEAT & GRILLED
CHICKEN SUMMER SALAD
(PAGE 64)

Buckwheat & grilled chicken summer salad

PREPARATION TIME: 20 MINUTES
(+ 30 MINUTES MARINATING)
COOKING TIME: 25 MINUTES
SERVES 4

60 ml (2 fl oz/¼ cup) lemon juice

60 ml (2 fl oz/¼ cup) olive oil

2 garlic cloves, crushed

500 g (1 lb 2 oz) skinless chicken thigh fillets, fat trimmed

200 g (7 oz/1 cup) raw buckwheat

1 Lebanese (short) cucumber, trimmed and diced

250 g (9 oz) mixed cherry and grape tomatoes, halved or quartered depending on size

¼ cup flat-leaf (Italian) parsley leaves

¼ cup mint leaves, torn

¼ cup basil leaves, torn

4 spring onions (scallions), trimmed and thinly sliced

1 teaspoon sumac (see tips), plus extra, to garnish

** This flavoursome salad is similar to a tabouleh, but unlike the traditional cracked wheat version it is gluten free.*

1 Put half the lemon juice, half the oil and the garlic in a shallow glass or ceramic container and stir to combine. Add the chicken and stir to coat. Cover and set aside for 30 minutes to marinate.

2 Meanwhile, heat a large non-stick frying pan over medium–high heat. Add the buckwheat and cook, stirring, for 3–4 minutes or until fragrant. Put the toasted buckwheat and 500 ml (17 fl oz/2 cups) water in a saucepan and bring to the boil. Cover, reduce the heat to low and simmer for 10–12 minutes or until just tender (be careful not to overcook). Rinse under cold running water, then drain well.

3 Transfer the buckwheat to a large mixing bowl and add the cucumber, tomatoes, herbs, spring onions, sumac and the remaining lemon juice and olive oil. Stir to combine, then season with sea salt and freshly ground black pepper.

4 Preheat a chargrill pan over medium–high heat. Cook the chicken for 4 minutes each side or until lightly charred and cooked through. Set aside to cool slightly, then thinly slice. Add the chicken to the salad and toss to combine. Serve sprinkled with a little extra sumac.

TIPS Sumac is a spice used in Middle Eastern and Mediterranean cooking. It has a tangy citrus flavour and a vibrant deep-red colour.

Toasting the buckwheat before cooking only takes a few minutes, but it results in a much better flavour so it is worth the effort. I find toasting it myself is better than buying commercial toasted buckwheat (known as kasha), as it can be very dark in colour and overpowering in taste.

Buckwheat risotto with lemon & garlic prawns

PREPARATION TIME: 20 MINUTES
COOKING TIME: 40 MINUTES
SERVES 4

1 tablespoon lemon juice
2 garlic cloves, crushed
2 teaspoons finely grated lemon zest
2 tablespoons olive oil
12 raw large prawns (shrimp), peeled and deveined with tails left intact
1 litre (35 fl oz/4 cups) chicken stock (see tip)
300 g (10½ oz/1½ cups) raw buckwheat
1 tablespoon unsalted butter
1 large white onion, finely chopped
1 medium fennel bulb, trimmed and finely chopped
125 ml (4 fl oz/½ cup) white wine
2 tablespoons thin (pouring/whipping) cream
35 g (1¼ oz/⅓ cup) finely grated parmesan cheese
100 g (3½ oz) baby spinach leaves

* *When cooked in a risotto style, buckwheat is surprisingly creamy and has a delicious nutty flavour. It takes about the same time to cook as arborio rice, but has over three times as much dietary fibre.*

1 Put the lemon juice, half the garlic, half the lemon zest and half the olive oil in a shallow glass or ceramic bowl. Add the prawns and stir until they are coated in the garlic mixture. Cover and set aside to marinate while you make the risotto.

2 Bring the chicken stock to the boil in a medium saucepan over high heat. Reduce the heat to very low and keep at a gentle simmer. Meanwhile, heat a large non-stick frying pan over medium–high heat and cook the buckwheat, stirring, for 3–4 minutes or until fragrant. Remove from the heat and set aside.

3 Heat the butter and remaining olive oil in a large heavy-based saucepan over medium heat. Cook the onion and fennel, stirring, for 6–7 minutes or until softened. Add the remaining garlic and cook for 1 minute more. Add the toasted buckwheat and stir until the grains are well coated in the butter mixture.

4 Add the wine and simmer until it has almost evaporated. Start adding the stock, a ladleful at a time. Cook, stirring, until almost all the stock has evaporated before adding another ladleful.

5 When you have added all the stock and the buckwheat is *al dente* and creamy (this will take about 25 minutes), add the remaining lemon zest, the cream and parmesan. Season with salt and pepper and stir to combine, then remove from the heat. Cover and stand for 5 minutes, then stir through the spinach leaves.

6 Meanwhile, heat a large non-stick frying pan over high heat. Cook the prawns for 1–2 minutes each side or until they are golden and just cooked through. To serve, divide the risotto among serving bowls and top each with 3 prawns.

TIP If you want this dish to be gluten free, make sure you use gluten-free stock.

Buckwheat, corn & chive fritters with crisp bacon

PREPARATION TIME: 20 MINUTES
COOKING TIME: 20 MINUTES
SERVES 4

185 ml (6 fl oz/3/4 cup) milk
2 eggs, separated
75 g (2 3/4 oz/1/2 cup) plain (all-purpose) flour
65 g (2 1/4 oz/1/2 cup) buckwheat flour
2 teaspoons baking powder
400 g (14 oz/2 cups) fresh corn kernels
1 zucchini (courgette), coarsely grated, squeezed of excess moisture
50 g (1 3/4 oz/1/2 cup) coarsely grated cheddar cheese (see tips)
2 tablespoons snipped chives
2 tablespoons olive oil
Grilled (broiled) bacon, chopped avocado and baby spinach leaves, to serve

** These flavoursome fritters make a wonderful brunch or light lunch.*

1 Whisk together the milk and egg yolks in a large bowl. Gradually whisk in the combined flours and baking powder until smooth and well incorporated. Stir through the corn, zucchini, cheese and chives.

2 Use an electric mixer with a whisk attachment to whisk the egg whites in a clean, dry bowl until stiff peaks form. Gently fold the egg whites into the batter until just combined.

3 Heat half the oil in a large non-stick frying pan over medium–high heat. Spoon 80 ml (2 1/2 fl oz/1/3 cup) of mixture per fritter into the pan, to cook 2–3 fritters at a time. Cook for 3 minutes each side or until golden and cooked through. Transfer to a plate, cover loosely with foil and keep warm. Continue with the remaining batter to make 8 fritters, adding a little more oil to the pan as necessary.

4 Serve the fritters with grilled bacon, avocado and baby spinach.

TIPS You can substitute any type of cheese for the grated cheddar, such as crumbled feta, ricotta or even goat's cheese.

Whisking the egg whites and adding them to the batter at the end results in a much lighter, less dense fritter.

Buckwheat pikelets with blueberries & sweet ricotta

PREPARATION TIME: 15 MINUTES
COOKING TIME: 10 MINUTES
MAKES ABOUT 22

90 g (3¼ oz/⅔ cup) buckwheat flour
75 g (2¾ oz/½ cup) plain (all-purpose) flour
1½ teaspoons baking powder
2 tablespoons caster (superfine) sugar
2 eggs, separated
200 ml (7 fl oz) buttermilk
2 teaspoons natural vanilla extract
250 g (9 oz/1 cup) fresh ricotta cheese
1 tablespoon icing (confectioners') sugar mixture, sifted
Unsalted butter, for cooking
120 g (4¼ oz) punnet fresh blueberries

* *Buckwheat flour has quite a strong flavour, so when using it to make pikelets or pancakes it is best to use some regular wheat flour as well.*

1 Combine the flours, baking powder and caster sugar in a large bowl. Make a well in the centre, add the egg yolks, buttermilk and half the vanilla and whisk until smooth. The batter will be quite stiff.

2 Use an electric mixer with a whisk attachment to whisk the egg whites in a clean, dry bowl until stiff peaks form. Gently fold the egg whites into the batter until just combined. The batter will be quite sticky now.

3 To make the sweet ricotta, put the ricotta, remaining vanilla and icing sugar in a bowl and stir to combine. Set aside.

4 Heat a little butter in a non-stick frying pan over medium–high heat. Drop tablespoonfuls of the mixture into the pan and cook for 1–2 minutes each side or until the pikelets are golden and cooked through. Transfer to a plate, cover loosely with foil and keep warm. Repeat with the remaining mixture, adding a little more butter to the pan as necessary.

5 Serve the pikelets topped with a teaspoon of sweet ricotta and a few blueberries.

TIP For a savoury variation to serve as a canape, omit the sugar and serve topped with a little cream cheese and some thinly sliced smoked salmon.

Buckwheat, honey & nut muesli bars

PREPARATION TIME: 15 MINUTES
COOKING TIME: 45–55 MINUTES
MAKES 16

235 g (8½ oz/⅔ cup) honey
125 ml (4 fl oz/½ cup) sunflower oil
60 g (2¼ oz/¼ cup, firmly packed) light brown sugar
250 g (9 oz/2½ cups) rolled (porridge) oats
10 g (¼ oz/1 cup) puffed buckwheat
45 g (1¾ oz/½ cup) desiccated coconut
1 teaspoon ground cinnamon
75 g (2¾ oz/½ cup) sweetened dried cranberries
85 g (3 oz/½ cup) coarsely chopped seedless raisins
80 g (2¾ oz/½ cup) raw cashew nuts, coarsely chopped

* *Home-made muesli bars are healthier and tastier than bought ones and you know exactly what is in them. I find that cooking them for a longer time at a lower temperature gives a crisp and golden result and you do not need to use as much oil or butter.*

1 Preheat the oven to 140°C (275°F/Gas 1). Line a 23 x 33 cm (9 x 13 inch) baking tin with non-stick baking paper.

2 Put the honey, oil and sugar in a medium saucepan over medium heat and cook, stirring, until the sugar has dissolved and the ingredients are well combined. Set aside.

3 Put the rolled oats, buckwheat, coconut, cinnamon, cranberries, raisins and nuts in a large bowl. Add the honey mixture and stir to combine. Using slightly wet hands, press the mixture firmly into the lined tin. Press the mixture with the back of a spoon to make the surface smooth and even.

4 Bake for 40–50 minutes or until the surface is dark golden brown all over. Cool completely in the tin before cutting into 16 bars.

TIPS For a gluten-free version, replace the oats with quinoa flakes.

You can substitute puffed millet or amaranth for the puffed buckwheat if you prefer.

These muesli bars will keep in an airtight container for up to 5 days. For extra-crisp muesli bars, store in an airtight container in the refrigerator.

Brown rice

NUTRITIONAL INFORMATION
(per 195 g/6¾ oz/1 cup cooked brown rice)

Energy: 913 kilojoules, 218 calories Protein: 4.5 g Total fat: 1.6 g
Saturated fat: 0.3 g Carbohydrate: 45.8 g Sugars: 1 g Dietary fibre: 3.5 g
Cholesterol: 0 mg Vitamin B_6: 0.3 mg Iron: 1.0 mg Magnesium: 84 mg
Manganese: 2.1 mg Phosphorus: 150 mg Sodium: 2 mg

Rice was first cultivated in China, with records going back to about 2500BC. Archaeological discoveries show that rice was even grown long before this, but in dry conditions rather than rice paddies. For the majority of its long history rice was a staple only in Asia, until traders and conquerors introduced it to all corners of the world. The majority of rice is still grown in Asia, where it is an important part of the food culture of those countries. Thailand, Vietnam and China are the three largest exporters of rice.

In Asia, brown rice is often referred to as 'rough' rice and it was traditionally rarely eaten except by the poor, the sick and the elderly. Nowadays the tables have turned and brown rice is valued for its higher nutritional value and is enjoying an emerging popularity.

Brown rice has a delicious nutty taste, slightly chewy texture and a far superior nutrient profile than highly processed white rice. This is because brown rice is the whole grain, with only the inedible outer husk removed, and therefore it retains the nutrient-rich bran and germ. White rice is both milled and polished, which removes the bran layer and germ, and the nutrients they contain, particularly vitamins B_2 and B_3, iron and magnesium. Brown rice and white rice do have similar calorific (kilojoule) and carbohydrate values, however.

Brown rice is a great source of carbohydrate — in fact, 90 per cent of the kilojoule content of rice is due to carbohydrate. As brown rice is not as refined as white rice, it takes longer to digest and contains significantly more fibre, making you feel satisfied faster. Brown rice is an excellent source of the trace mineral manganese, and the minerals magnesium and selenium. One cup of cooked brown rice contains 84 mg magnesium, compared to just 19 mg in white rice.

Brown rice also contains phytochemicals, including phenolic acids, phytic acid, plant sterols and saponins. The bran in brown rice is high in insoluble dietary fibre.

Brown rice is most readily available in long-grain and medium-grain varieties. Long-grain brown rice is ideal to use in pilafs and warm or cold salads, where you want the grains to remain fluffy and separated. Medium-grain is a great all-purpose rice with a little more starch, so it can be used in stuffings for meat and vegetables, salads and even to make sushi. Short-grain is ideal for rice puddings and risotto, where you want a rich creamy texture. Brown rice risotto takes over an hour to cook, so I find baking it is the easiest method as it does not require constant stirring.

Brown rice flakes are ideal for making gluten-free porridge, they only take minutes to cook, so are great for a quick breakfast. Also, try adding rice bran flakes to home-made muesli.

To cook brown rice

To cook by absorption, place 200 g (7 oz/1 cup) long- or medium-grain brown rice in a saucepan with 375 ml (13 fl oz/1½ cups) water and bring to the boil. Reduce the heat and simmer, covered, for 25–30 minutes or until the water has evaporated. Remove from the heat, keep covered and set aside for 5 minutes.

To boil, place the same amount of rice in a large saucepan, add 1.5 litres (52 fl oz/6 cups) water and bring to the boil, stirring occasionally. Reduce the heat and boil gently, uncovered, for 25–30 minutes. Drain well.

Fried rice with Chinese sausage, egg & kecap manis

PREPARATION TIME: 15 MINUTES
COOKING TIME: 10 MINUTES
SERVES 4

1½ tablespoons sunflower oil
3 eggs, lightly whisked
1 white onion, finely chopped
2 celery stalks, diced
175 g (6 oz) Chinese sausage (lap cheong, see tip), thinly sliced
1 long red chilli, seeded and finely chopped
2 teaspoons finely grated ginger
780 g (1 lb 11½ oz/4 cups) cold cooked medium-grain brown rice (see page 75)
130 g (4½ oz/1 cup) frozen green peas, thawed
2 tablespoons light soy sauce
1 tablespoon kecap manis
4 spring onions (scallions), trimmed and thinly sliced, to garnish

* *Fried rice is an excellent way to use up leftover rice. This version contains almost seven times the amount of fibre than one based on white rice.*

1 Heat 1 teaspoon of the oil in a large non-stick wok or frying pan over medium heat. Pour half the whisked egg into the wok and swirl to cover the base. Cook for 1 minute or until set. Carefully loosen the edges, turn out onto a board and set aside to cool. Repeat with another teaspoon of oil and the remaining egg. Roll up both the omelettes and cut into thin strips. Set aside.

2 Heat the remaining oil in the wok over high heat. Add the onion and celery and stir-fry for 2 minutes, then add the Chinese sausage and stir-fry for a further 2 minutes or until crisp. Add the chilli and ginger and stir-fry for 30 seconds. Add the rice and peas and cook, tossing, for 2–3 minutes or until the rice is heated through. Add the soy sauce and kecap manis and toss to combine.

3 Serve the fried rice garnished with the sliced omelette and spring onions.

TIP Chinese sausage is a spiced dried pork sausage that can be found in the Asian section of supermarkets or from Asian grocers.

Lamb biryani with coconut & chilli relish

PREPARATION TIME: 20 MINUTES
(+ 2–3 HOURS MARINATING)
COOKING TIME: 1 HOUR 30 MINUTES
SERVES 4

1 tablespoon ground cumin

1 tablespoon sweet paprika

1 teaspoon ground turmeric

70 g (2½ oz/¼ cup) plain yoghurt

700 g (1 lb 9 oz) diced boneless lamb leg

300 g (10½ oz/1½ cups) brown basmati (or long-grain) rice

2 tablespoons sunflower oil

2 brown onions, thinly sliced

2 garlic cloves, crushed

2 teaspoons finely grated ginger

875 ml (30 fl oz/3½ cups) chicken stock (see tip)

Baby spinach leaves, plain yoghurt and lime wedges (optional), to serve

COCONUT & CHILLI RELISH

2 long green chillies, seeded and finely chopped

¼ cup coarsely chopped coriander (cilantro) leaves

35 g (1¼ oz/½ cup) shredded coconut, lightly toasted

2 tablespoons lime juice

½ teaspoon caster (superfine) sugar

1 Put half each of the cumin, paprika and turmeric in a shallow glass or ceramic dish. Add the yoghurt and stir to combine. Add the lamb and stir to coat, then cover with plastic wrap and place in the refrigerator for at least 2–3 hours to marinate. Soak the brown rice in a bowl of cold water for 1 hour, then rinse well in a sieve and drain.

2 Meanwhile, preheat the oven to 180°C (350°F/Gas 4). Heat 2 teaspoons of the oil in a large flameproof casserole dish over high heat. Cook the lamb, in batches, for 2–3 minutes or until golden, adding another 2 teaspoons of oil when necessary. Transfer to a bowl and set aside.

3 Reduce the heat to medium, add the remaining oil and the onions to the casserole dish and cook, stirring occasionally, for 5 minutes. Add the garlic, ginger and remaining spices and cook, stirring, for 2 minutes or until fragrant. Add the rice, stirring to coat in the spices. Add the stock and bring to the boil, then return the lamb to the dish in a single layer on top of the rice.

4 Cover the dish with a lid and bake for 1 hour 15 minutes or until the stock is absorbed and the lamb and rice are tender.

5 Meanwhile, to make the coconut & chilli relish, combine all the ingredients in a medium bowl and season with sea salt and freshly ground black pepper, to taste.

6 Serve the biryani with the coconut & chilli relish, baby spinach, yoghurt and a wedge of lime, if desired.

TIP If you want this dish to be gluten free, make sure you use gluten-free stock.

gluten-free

Rice salad with roasted pumpkin, beans & orange spice dressing

PREPARATION TIME: 25 MINUTES
COOKING TIME: 30 MINUTES
SERVES 4

800 g (1 lb 12 oz) pumpkin (winter squash), peeled, seeded and cut into 1.5 cm (5/8 inch) cubes

1 tablespoon olive oil

250 g (9 oz) green beans, trimmed and cut into 3 cm (1 1/4 inch) lengths

585 g (1 lb 4 1/2 oz/3 cups) cooked medium-grain brown rice (see page 75), cooled

1 zucchini (courgette), trimmed, halved and thinly sliced

40 g (1 1/2 oz/1/4 cup) sunflower seeds

1/4 cup flat-leaf (Italian) parsley leaves, coarsely chopped

1/4 cup mint leaves, shredded

ORANGE SPICE DRESSING

80 ml (2 1/2 fl oz/1/3 cup) freshly squeezed orange juice

2 tablespoons olive oil

3 teaspoons white wine vinegar

1 teaspoon honey

1/2 teaspoon ground cinnamon

1 1/2 teaspoons ground cumin

I love using brown rice in salads as it gives so much flavour and texture. It stays al dente *even after the dressing is added.*

1 Preheat the oven to 200°C (400°F/Gas 6). Put the pumpkin on a large baking tray lined with non-stick baking paper. Season with sea salt and freshly ground black pepper and drizzle with the olive oil. Roast for 25–30 minutes, until the pumpkin is golden and tender.

2 Meanwhile, blanch the beans in a saucepan of boiling water until bright green and tender-crisp. Refresh under cold running water, then drain well.

3 Put the roasted pumpkin, beans, rice, zucchini, sunflower seeds, parsley and mint in a large bowl and stir to combine.

4 To make the dressing, put all the ingredients in a small bowl and whisk to combine. Add to the salad and stir to combine. Season with sea salt and freshly ground black pepper, to taste.

TIP This salad will keep, covered, in the refrigerator for up to 2 days.

SUSHI WITH SPICY SESAME CHICKEN & AVOCADO (PAGE 84)

gluten-free

Sushi with spicy sesame chicken & avocado

PREPARATION TIME: 30 MINUTES (+ 1 HOUR MARINATING AND 30 MINUTES CHILLING)
COOKING TIME: 45 MINUTES
MAKES 4 LARGE ROLLS

300 g (10½ oz/1½ cups) medium-grain or short-grain brown rice

60 ml (2 fl oz/¼ cup) seasoned rice vinegar

60 ml (2 fl oz/¼ cup) white vinegar

4 sheets nori

1 Lebanese (short) cucumber, cut into 5 cm (2 inch) batons

½ avocado, peeled, stone removed and cut into 5 cm (2 inch) batons

Pickled ginger, wasabi and tamari (see tip), to serve

SPICY SESAME CHICKEN

¼ teaspoon dried red chilli flakes

1 garlic clove, crushed

2 tablespoons tamari (see tip)

1 tablespoon sesame oil

2 (about 250 g/9 oz in total) skinless chicken thigh fillets

** Brown rice takes longer to break down in your body than white rice, so this sushi will help your blood sugar levels remain stable throughout the day.*

1 To make the spicy sesame chicken, put the chilli flakes, garlic, tamari and sesame oil in a shallow glass or ceramic container. Add the chicken and stir to coat, then cover and put in the refrigerator for at least 1 hour to marinate.

2 Meanwhile, put the rice in a large saucepan with 750 ml (26 fl oz/3 cups) cold water and bring to the boil over high heat. Reduce the heat to low, cover and simmer for 30–35 minutes, until all the water is absorbed. Remove from the heat, cover and set aside for 5 minutes.

3 Add the seasoned rice vinegar to the rice and stir until well combined. Cover a large baking tray with foil, then spread the rice evenly over the tray and set aside to cool completely.

4 Meanwhile, heat a chargrill pan or non-stick frying pan over high heat and cook the chicken for 3–4 minutes each side, until golden and cooked through. Remove and set aside to cool, then cut into 1 cm (½ inch) thick slices.

5 Combine the white vinegar and 250 ml (9 fl oz/1 cup) cold water in a small bowl. Divide the rice into 4 portions. Place a sheet of nori, shiny side down, on a bamboo mat. Dip your hands in the vinegar mixture, then use them to spread one portion of the rice evenly over the bottom two-thirds of the nori sheet, leaving a small border around the edge.

6 Place a few slices of chicken along the middle of the rice and top with some cucumber and avocado. Lift up the end of the mat closest to you and roll it over the ingredients to enclose, then keep rolling to make a complete roll. Continue with the remaining nori, rice and fillings. Wrap each roll tightly in plastic wrap and refrigerate for 30 minutes. Slice and serve with pickled ginger, wasabi and tamari.

TIP If you want this dish to be gluten free, make sure you use gluten-free tamari.

Brown rice porridge with vanilla baked rhubarb

PREPARATION TIME: 10 MINUTES
COOKING TIME: 15 MINUTES
SERVES 4

375 ml (13 fl oz/1½ cups) milk (see tips)

½ teaspoon ground cinnamon

180 g (6¼ oz/2 cups) brown rice flakes

VANILLA BAKED RHUBARB

600 g (1 lb 5 oz) rhubarb, trimmed and cut into 5 cm (2 inch) lengths

55 g (2 oz/¼ cup) caster (superfine) sugar

125 ml (4 fl oz/½ cup) freshly squeezed orange juice

1 teaspoon natural vanilla extract

** Compared to regular rice or brown rice, brown rice flakes take very little time to cook, so they are ideal for a quick breakfast. They are also a gluten-free alternative to traditional oat-based porridge and have a rich and creamy texture once cooked.*

1 Preheat the oven to 180°C (350°F/Gas 4). To make the vanilla baked rhubarb, place the rhubarb in a single layer in a roasting pan, sprinkle with the sugar and drizzle with the orange juice and vanilla. Cover the pan with foil and bake for 10–15 minutes (check the rhubarb after 10 minutes), until tender.

2 Meanwhile, put the milk, 500 ml (17 fl oz/2 cups) water and the cinnamon in a large saucepan. Bring to the boil over medium heat.

3 Add the brown rice flakes to the pan and return to the boil, then reduce the heat to low and simmer, stirring, for 3–4 minutes or until thick and creamy. Serve topped with the rhubarb and any pan juices.

TIPS You can use any type of milk, such as soy or rice, in this recipe.

When rhubarb is out of season, you can try serving the porridge topped with fresh berries or sliced banana and a drizzle of maple syrup or honey.

Roasted pork loin with cumin, quince & pistachio stuffing

PREPARATION TIME: 25 MINUTES
(+ COOLING, AND 15 MINUTES RESTING)
COOKING TIME: 1 HOUR
SERVES 8

2 kg (4 lb 8 oz) rolled boneless pork loin (see tips)

1 tablespoon olive oil

2 teaspoons sea salt

Steamed green beans, to serve

STUFFING

1 tablespoon olive oil

1 large brown onion, finely chopped

1 teaspoon ground cumin

1 teaspoon ground coriander

390 g (13¾ oz/2 cups) cooked medium-grain brown rice (see page 75), cooled

100 g (3½ oz) quince paste, chopped

45 g (1¾ oz/⅓ cup) pistachios, chopped

2 tablespoons chopped flat-leaf (Italian) parsley

1 Preheat the oven to 220°C (425°F/Gas 7). To make the stuffing, heat the olive oil in a frying pan over medium heat and cook the onion, stirring occasionally, for 3–4 minutes or until softened. Add the spices and cook, stirring, for a further minute. Set aside.

2 Put the rice, onion mixture, quince paste, pistachios and parsley in a large bowl. Use clean hands to mix until combined (see tips). Season well with salt and pepper, then set aside to cool completely.

3 Unroll the pork loin and place, rind side up, on a board or clean work surface. Use a sharp knife to score the rind crossways at 2 cm (¾ inch) intervals. Turn the pork over and cut a vertical slit into the thick end of the fillet, going halfway through the meat. Pack the stuffing into the slit and along the centre of the loin.

4 Starting at the thicker end, tightly roll up the pork to enclose the filling. Tie with kitchen string at 2 cm (¾ inch) intervals, then place the pork rind-side up in a large roasting pan. Drizzle with the oil and rub the salt into the rind.

5 Roast the pork for 25 minutes or until the rind crackles, then reduce the temperature to 180°C (350°F/Gas 4) and roast for a further 45–50 minutes or until the rind is crisp and golden and a meat thermometer inserted into the thickest part of the meat reads 70°C (158°F). Transfer the pork to a plate (don't cover with foil) and set aside for 15 minutes to rest before slicing. Serve with green beans.

** Whole grains such as brown rice are great to use in stuffings, as they absorb all the juices without becoming soggy. Middle Eastern flavours such as cumin, quince and pistachio complement the nutty flavour of the rice.*

TIPS For extra crisp and crunchy crackling, place the pork loin on a baking tray, skin side up and uncovered, in the refrigerator overnight. You can ask your butcher to score the pork for you.

It is important to rub the quince paste evenly through the rice mixture, as it helps to make the rice stick together and this in turn makes it easier to stuff the pork.

You could use cooked pearl barley, buckwheat, millet or quinoa instead of the rice.

gluten-free

Cauliflower, rice & lentil salad with tahini dressing

PREPARATION TIME: 15 MINUTES
COOKING TIME: 15 MINUTES
SERVES 4

600 g (1 lb 5 oz) cauliflower, trimmed and cut into florets

2 tablespoons olive oil

1 brown onion, thinly sliced

2 teaspoons brown mustard seeds

2 teaspoons ground cumin

2 tablespoons currants

390 g (13¾ oz/2 cups) freshly cooked brown rice (see page 75)

400 g (14 oz) tin brown lentils, drained and rinsed

1 tablespoon lemon juice

2 tablespoons chopped coriander (cilantro) leaves, to garnish (optional)

TAHINI DRESSING

70 g (2½ oz/¼ cup) Greek-style yoghurt

1 tablespoon tahini (see tip)

1 tablespoon lemon juice

1 To make the tahini dressing, combine the yoghurt, tahini and lemon juice in a small bowl. Mix in 1–2 tablespoons warm water, enough to make a dressing-like consistency. Set aside.

2 Blanch the cauliflower in a saucepan of boiling water for 3–4 minutes or until tender-crisp. Refresh under cold running water, then drain well.

3 Heat the oil in a large deep frying pan over medium–high heat and cook the onion, stirring occasionally, for 3–4 minutes or until golden. Add the mustard seeds and cumin and cook, stirring, for 1 minute. Add the cauliflower and cook, stirring often, for a further 2–3 minutes, until the cauliflower is golden. Add a little extra oil to the pan if necessary.

4 Add the currants, rice and lentils to the pan and cook for 2 minutes or until heated through. Season with salt and freshly ground black pepper, then stir through the lemon juice. Serve the salad warm, drizzled with the tahini dressing and garnished with coriander, if desired.

TIP Tahini is a Middle Eastern paste made from ground sesame seeds and olive oil. You can find it in the health food section of your supermarket. Check the label to make sure it is gluten free.

** Whole grains and legumes are made for each other, as together they make a complete protein, which is especially important for vegetarians and vegans.*

Five-spice duck breast with ginger & spring onion rice

PREPARATION TIME: 20 MINUTES
COOKING TIME: 20 MINUTES
SERVES 4

4 x 150 g (5½ oz) duck breast fillets, with skin

1 tablespoon light brown sugar

½ teaspoon sea salt

1 teaspoon Chinese five-spice

100 g (3½ oz) fresh shiitake mushrooms, sliced

3 cm (1¼ inch) piece ginger, peeled and julienned

6 spring onions (scallions), trimmed and thinly sliced diagonally, plus extra, to garnish

200 g (7 oz) snow peas (mangetout), trimmed and thinly sliced

780 g (1 lb 11½ oz/4 cups) cooked brown rice (see page 75), cooled

2 tablespoons mirin

1 tablespoon tamari (see tip)

2 teaspoons sesame oil

** This dish is quick to make and tastes amazing! Brown rice pairs well with the robust flavours of shiitake mushrooms and duck, and using the rendered duck fat to stir-fry the rice adds even more flavour to the dish.*

1 Preheat the oven to 180°C (350°F/Gas 4). Use a sharp knife to score the skin of each duck breast 5–6 times. Combine the sugar, salt and five-spice. Rub the sugar mixture evenly into the skin of each duck breast.

2 Heat a large frying pan over medium heat and cook the duck, skin side down, for 6–7 minutes, until the skin is crisp and the fat has rendered down. Turn and cook for a further minute. Transfer to a large baking tray, skin side up, reserving 1 tablespoon of the rendered duck fat. Roast the duck for 5 minutes for medium, then transfer to a clean board. Cover loosely with foil and set aside for 5 minutes to rest.

3 Meanwhile, heat the reserved duck fat in a large wok over high heat. Add the mushrooms and stir-fry for 2 minutes, then add the ginger and spring onions and stir-fry for 1 minute more. Add the snow peas and stir-fry for 1 minute, then add the rice, mirin, tamari and sesame oil and stir-fry for a further 2 minutes or until the rice is heated through.

4 Thinly slice the duck. Divide the rice among 4 serving bowls, top with slices of duck and garnish with the extra spring onion.

TIP If you want this dish to be gluten free, make sure you use gluten-free tamari.

gluten-free

Creamy coconut rice with caramelised strawberries

PREPARATION TIME: 10 MINUTES
COOKING TIME: 30 MINUTES
SERVES 4

585 g (1 lb 4½ oz/3 cups) cooked medium-grain brown rice (see page 75)

400 ml (14 fl oz) tin low-fat coconut milk

185 ml (6 fl oz/¾ cup) milk

60 g (2¼ oz) grated palm sugar (jaggery)

1 cinnamon stick

CARAMELISED STRAWBERRIES

250 g (9 oz) punnet strawberries, hulled and halved

1½ tablespoons caster (superfine) sugar

** Although rice pudding is traditionally made with white rice, this brown-rice version has a delicious nutty flavour and is far higher in fibre.*

1 Put the cooked rice, coconut milk, milk, sugar and cinnamon stick in a large saucepan over medium heat. Bring to the boil, then reduce the heat to low and simmer, stirring occasionally, for 20–25 minutes or until thick and creamy. Set aside to cool slightly (see tips).

2 When you are ready to serve, make the caramelised strawberries. Heat a large frying pan over medium–high heat and add the strawberries and 2 teaspoons of water. Sprinkle the sugar over the strawberries. Cook, stirring occasionally, for 1–2 minutes or until the strawberries are slightly caramelised.

3 To serve, divide the rice pudding among serving glasses or bowls, and top with some caramelised strawberries and any pan juices.

TIPS This rice pudding can be served warm or at room temperature.

The trick to making a creamy brown-rice pudding is to cook the rice first, otherwise it can take over an hour of stirring!

For a vegan version, use soy, almond or rice milk.

Chia

NUTRITIONAL INFORMATION
(per 15 g/1/2 oz chia seeds)

Energy: 287 kilojoules, 68 calories **Protein:** 3.1 g **Total fat:** 4.9 g

Saturated fat: 0.5 g **Omega 3 ALA:** 3.2 g **Omega 6 LA:** 0.9 g

Carbohydrate: 0.2 g **Sugars:** <1 g **Dietary fibre:** 6.2 g **Calcium:** 107 g

Iron: 2.5 mg **Magnesium:** 59 mg **Phosphorus:** 160 mg **Sodium:** 0.3 mg

Chia is an annual herb belonging to the mint family, and is still grown today for its edible seeds. Chia is technically not a grain (like amaranth, quinoa and buckwheat), but deserves its place in this book due to its ancient history, recent revival and more importantly its unique and amazing nutritional properties.

Originally from Southern Mexico and Guatemala, chia seeds were an important food source for the Aztecs and Mayans. During this time chia was not only eaten as a grain, but used in beverages, medicines, pressed for its oil and ground to be used in breads.

Chia seeds are tiny in appearance, similar in size to sesame or poppy seeds. They are small flat discs with a slightly mottled appearance. They vary in colour from black to grey, brown and even white.

The most impressive quality of chia seeds is their omega 3 content. Chia seeds have the highest plant content of omega-3 fatty acids, which have a wide array of health benefits. These include cardiovascular health, cancer and diabetes prevention, relief from joint stiffness, assisting in depression treatment and even improved memory. Chia seeds are also a good source of omega-6 fatty acids.

Chia seeds are high in fibre, especially insoluble fibre, assisting in gastrointestinal health and reduced risk of colon cancer. Chia is also a rich protein source and contains all the essential amino acids, as well as several non-essential ones.

Chia is gluten free, so it is suitable for those with coeliac disease, and it is also rich in several vitamins and minerals including calcium, iron, magnesium, phosphorus and potassium.

Chia is available as seeds and, less commonly, as chia oil or bran. Chia seeds have a variety of food uses — they can be eaten raw, sprinkled on salads, dips, roasted vegetables or over muesli. They can be baked in muffins, muesli bars, biscuits, cakes and breads, or added to drinks such as smoothies. Chia seeds are mixed with water or fruit juice to make 'chia fresca' in Mexico.

Chia oil is produced by extracting the oil from chia seeds, and is a wonderful plant source of omega 3. The oil should not be heated as it will lose its nutritional value. It has a mild flavour, so is ideal to drizzle over salads or combine with olive oil in salad dressings. You can also add a teaspoon of the oil to a smoothie. Once opened, chia oil needs to be kept in the fridge.

Chia bran is the husk of the chia seed, and is extremely high in dietary fibre, containing both soluble and insoluble fibre, which assists with overall digestive health. It can be added to baked items such as muffins, to smoothies to give body or sprinkled over cereal.

Chia's versatility and ease of preparation make it extremely easy to incorporate into your diet. You also only need a relatively small serve to gain nutritional benefits.

To make chia gel

Add 60 g (2¼ oz/⅓ cup) chia seeds to 500 ml (17 fl oz/ 2 cups) cold water in a large airtight container with a lid. Cover and shake the container until well combined, then set aside for 20–30 minutes or until a gel forms, shaking the container every 10 minutes. This mixture can keep, covered, in the fridge, for up to 2 weeks. Add it to drinks to give them a nutritional boost.

gluten-free

Chia, mango & raspberry breakfast smoothie

PREPARATION TIME: 5 MINUTES
COOKING TIME: NIL
SERVES 2

350 ml (12 fl oz) milk

130 g (4½ oz/½ cup) plain yoghurt

1½ tablespoons white chia seeds

1 mango, seed and skin removed, flesh diced

60 g (2¼ oz/½ cup) fresh or frozen raspberries

2 teaspoons honey

✻ Putting chia seeds in your breakfast smoothie adds a massive 4.2 g (⅛ oz) of fibre per serve, as well as a good dose of omega-3 and omega-6 fatty acids.

1 Put all the ingredients in a blender and blend until smooth. Serve immediately.

TIP If you prefer a smoother texture, you can make a gel from the chia seeds before adding to your smoothie. Put the chia seeds in a jug with 250 ml (9 fl oz/1 cup) of water and set aside for 10 minutes or until a gel forms. Reduce the milk to 200 ml (7 fl oz) and add the chia gel with the remaining ingredients.

Roasted root vegetables with chia dukkah

PREPARATION TIME: 25 MINUTES
COOKING TIME: 45 MINUTES
SERVES 4

6 medium beetroot (beets) (about 700 g/1 lb 9 oz in total), peeled and cut into wedges

2 large parsnips (about 400 g/14 oz in total), peeled and cut into thick batons

1 large bunch Dutch (baby) carrots (about 500 g/1 lb 2 oz in total), trimmed, scrubbed and halved lengthways if large

2 tablespoons olive oil, plus extra, to drizzle

CHIA DUKKAH

1 tablespoon coriander seeds

1 tablespoon cumin seeds

70 g (2½ oz/½ cup) pistachios, lightly toasted

2 tablespoons white chia seeds

2 teaspoons sea salt

** Chia seeds work just like sesame seeds in this delicious dukkah. They also add dietary fibre, essential fatty acids and protein. Just add a leafy salad, goat's cheese and a slice of bread to make this a complete meal, or serve as a warm vegetable side.*

1 Preheat the oven to 200°C (400°F/Gas 6). Put the vegetables in a single layer on a large baking tray lined with non-stick baking paper. Drizzle with the olive oil and season with sea salt and freshly ground black pepper. Roast the vegetables for 40–45 minutes or until tender.

2 Meanwhile, to make the chia dukkah, put the coriander and cumin seeds in a small dry frying pan over medium–high heat. Cook, stirring, for 2–3 minutes or until fragrant. Transfer to a mortar and pestle and pound until finely crushed.

3 Put the toasted spices, pistachios, chia seeds and salt in a small food processor and process until finely chopped.

4 To serve, pile the roasted vegetables onto a serving platter, sprinkle with the chia dukkah and drizzle with a little extra olive oil.

TIP The dukkah will keep, stored in an airtight container, for up to 1 month. Try sprinkling it over grilled (broiled) meats (such as lamb cutlets, chicken or beef) or grilled pitta or Turkish bread as part of a mezze plate.

Smoked chicken, chia & broccolini salad with chilli sesame dressing

PREPARATION TIME: 15 MINUTES
COOKING TIME: 5 MINUTES
SERVES 4

2 bunches broccolini, trimmed and cut into long florets

2 x 200 g (7 oz) smoked chicken breast fillets (see tips), skin removed and thinly sliced

1 firm ripe avocado, peeled, stone removed and cut into thin wedges

100 g (3½ oz) baby spinach leaves

1 tablespoon black chia seeds

DRESSING

2 teaspoons sesame oil

1½ tablespoons mirin

1 tablespoon olive oil

1 tablespoon tamari (see tips)

1 tablespoon rice wine vinegar

1 long red chilli, seeded and finely chopped

1 To make the dressing, put all the ingredients in a small bowl and whisk to combine. Set aside.

2 Blanch the broccolini in a large saucepan of boiling water until bright green and tender-crisp. Refresh under cold running water, then drain well.

3 Put the broccolini, chicken, avocado and spinach leaves in a large bowl and gently toss to combine.

4 Divide the salad among 4 serving bowls, drizzle with a little of the dressing and sprinkle with the chia seeds.

TIPS You can replace the smoked chicken breast with two 200 g (7 oz) grilled chicken breasts, thinly sliced, or 350 g (12 oz/2 cups) shredded skinless barbecued chicken.

If you want this dish to be gluten free, make sure you use gluten-free smoked chicken and tamari.

** A really simple way to include chia in your diet is to sprinkle the seeds over fresh salads. As chia does have gelling properties once added to liquid, this is best done just before serving.*

Chia-crusted salmon with Asian greens & tamari dressing

PREPARATION TIME: 10 MINUTES
COOKING TIME: 10 MINUTES
SERVES 4

2 tablespoons white chia seeds
2 tablespoons black chia seeds
4 x 150 g (5½ oz) skinless salmon fillets
2 bunches choy sum, washed and trimmed
2 tablespoons sunflower oil
3 cm (1¼ inch) piece ginger, peeled and julienned
2 garlic cloves, thinly sliced
Noodles (see tip) or steamed brown rice, to serve

TAMARI DRESSING
2 tablespoons oyster sauce (see tip)
2 tablespoons tamari (see tip)
1 tablespoon Chinese rice wine
1 teaspoon caster (superfine) sugar

** Chia seeds are a great alternative to breadcrumb crusts for those on a gluten-free diet.*

Chia seeds, the highest plant source of omega fatty acids, and salmon ensure this meal is packed with essential fatty acids. Asian greens are rich in vitamins A, C and K, and are a good source of potassium and calcium.

1 Combine the white and black chia seeds on a plate. Press each salmon fillet in the chia seeds to evenly coat one side, then set aside.

2 Remove the stems from the choy sum, cut in half if long and reserve. To make the tamari dressing, put all the ingredients in a small bowl and stir to dissolve the sugar.

3 Heat 1 tablespoon of the oil in a large non-stick frying pan over high heat. Cook the salmon, chia side down, for 2–3 minutes or until golden. Turn and cook for a further 2 minutes (for medium) or until cooked to your liking. Set aside and keep warm.

4 Meanwhile, heat the remaining oil in a large wok or frying pan over high heat. Add the ginger and garlic and stir-fry for 30 seconds. Add the choy sum stems and stir-fry for 1–2 minutes, then add the choy sum leaves and stir-fry for 1 minute more or until almost wilted. Add half the dressing and toss to combine.

5 To serve, divide the choy sum among serving plates, top each with a piece of salmon and drizzle over a little of the remaining dressing. Serve with noodles or steamed rice.

TIP If you want this dish to be gluten free, make sure you use gluten-free noodles, oyster sauce and tamari.

gluten-free

Grilled asparagus with lemon & garlic dressing

PREPARATION TIME: 15 MINUTES
COOKING TIME: 5 MINUTES
SERVES 4

2 bunches asparagus, trimmed
1 tablespoon olive oil
1 tablespoon black chia seeds

DRESSING
1 tablespoon chia oil
1 tablespoon olive oil
1 tablespoon lemon juice
1 teaspoon finely grated lemon zest
1 garlic clove, crushed

1 To make the dressing, put all the ingredients in a small bowl and whisk to combine. Set aside.

2 Preheat a chargrill pan or barbecue grill plate on high. Drizzle the asparagus with the olive oil and grill for 2–3 minutes, turning, or until lightly charred and tender-crisp.

3 Transfer the asparagus to a serving plate, drizzle with the dressing and season with sea salt and freshly ground black pepper. Sprinkle with the chia seeds and serve immediately.

** Chia oil is an excellent plant source of omega-3 fatty acids, which have a wide array of health benefits. These include cardiovascular health, cancer and diabetes prevention, relief from joint stiffness, assisting in depression treatment and improved memory.*

Unlike other sources of omega-3 fatty acids, such as fish oil, chia oil has a mild flavour so is ideal to add to salad dressings.

CHIA CREPES WITH STAR ANISE MANDARINS (PAGE 110)

Chia crepes with star anise mandarins

PREPARATION TIME: 20 MINUTES
(+ 1 HOUR CHILLING)
COOKING TIME: 25 MINUTES
SERVES 4

110 g (3¾ oz/¾ cup) plain (all-purpose) flour

1 teaspoon caster (superfine) sugar

1 egg, lightly whisked

250 ml (9 fl oz/1 cup) milk

20 g (¾ oz) butter, melted, plus extra, for greasing

2 tablespoons black chia seeds

Greek-style yoghurt, to serve

STAR ANISE MANDARINS

150 g (5½ oz/⅔ cup) caster (superfine) sugar

1 cinnamon stick

2 star anise

3 seedless mandarins (see tips), peeled, white pith removed and segmented

* *Chia seeds don't just make these crepes look fantastic visually, they also add texture.*

1 Sift the flour into a large bowl and stir through the sugar and a pinch of salt. Add the egg, milk and melted butter and whisk until smooth. Refrigerate for at least 1 hour.

2 Meanwhile, to make the star anise mandarins, put the sugar, cinnamon stick and star anise in a medium saucepan with 160 ml (5¼ fl oz/⅔ cup) water. Cook, stirring, over medium heat to dissolve the sugar, then bring to the boil. Reduce the heat and simmer, without stirring, for 5 minutes or until syrupy. Add the mandarin segments and simmer for 2 minutes, then remove from the heat and set aside to cool in the syrup.

3 Heat a medium non-stick frying pan over medium heat and brush the pan with a little extra butter. Stir the chia seeds into the batter (see tips). Ladle 60 ml (2 fl oz/¼ cup) batter into the pan and tilt the pan slightly to quickly and evenly spread it out. Cook for 1 minute, then lift the outer edge and flip it over. Cook for a couple of seconds on the other side, then transfer to a plate and keep warm. Repeat with the remaining batter to make 8 crepes.

4 To serve, fold each crepe into quarters and place 2 crepes on each serving plate. Top with a dollop of yoghurt and spoon over some of the mandarins and syrup. Serve immediately.

TIPS You can replace the yoghurt with thick (double/heavy) cream or even mascarpone cheese for a more decadent dessert.

When mandarins are out of season, substitute 3 large navel oranges, peeled, white pith removed and segmented.

It is important to add the chia seeds to the batter just before you begin cooking the crepes, due to their gelling properties.

Chia seed Italian cookies

PREPARATION TIME: 20 MINUTES
COOKING TIME: 20 MINUTES
MAKES 36

2 egg whites, at room temperature

220 g (7¾ oz/1 cup) caster (superfine) sugar

100 g (3½ oz/1 cup) almond meal

100 g (3½ oz) milk chocolate (see tips), chopped

90 g (3¼ oz/½ cup) black chia seeds

75 g (2¾ oz/½ cup) sesame seeds

* *These more-ish little cookies are packed with fibre and good oils. They're the perfect thing to have with a coffee or cup of tea. Or, try crumbling them and layering in short glasses with cream and berries for a simple and delicious dessert.*

1 Preheat the oven to 140°C (275°F/Gas 1). Line 2 large baking trays with non-stick baking paper.

2 Use an electric mixer with a whisk attachment to whisk the egg whites in a large bowl until firm peaks form. With the motor running, gradually add 110 g (3¾ oz/½ cup) of the sugar, a spoonful at a time, whisking until it is dissolved. Combine the almond meal and the remaining sugar. Fold the almond mixture, chocolate, chia and sesame seeds through the egg white mixture.

3 Spoon 2 teaspoonfuls of the mixture per biscuit onto the lined trays, leaving 4 cm (1½ inches) between each to allow for spreading. Bake for 20 minutes, swapping the trays halfway through cooking, or until just golden and crisp. Remove and set aside on the trays to cool completely.

TIPS If you want these cookies to be gluten free, make sure you use gluten-free chocolate.

Keep in an airtight container for up to 1 week.

Chia, banana & chocolate muffins

PREPARATION TIME: 15 MINUTES
COOKING TIME: 20 MINUTES
MAKES 12

300 g (10½ oz/2 cups) self-raising flour
2 tablespoons black chia seeds
45 g (1¾ oz/½ cup) desiccated coconut
150 g (5½ oz/⅔ cup) caster (superfine) sugar
240 g (8¾ oz/1 cup) mashed banana
100 g (3½ oz) unsalted butter, melted and cooled
125 ml (4 fl oz/½ cup) milk
100 g (3½ oz/⅔ cup) chopped milk chocolate

** Add chia seeds to baked goods such as muffins, cakes and cookies as you would sesame seeds or poppy seeds for extra dietary fibre and omega-3 fatty acids.*

1 Preheat the oven to 180°C (350°F/Gas 4). Brush a 12-hole 80 ml (2½ fl oz/⅓ cup) muffin tin with melted butter to grease or line with paper cases.

2 Sift the flour into a large mixing bowl and stir through the chia seeds, coconut and sugar. Combine the banana, melted butter and milk, add to the dry ingredients and stir to combine — do not over-mix. Fold through the chopped chocolate.

3 Divide the mixture among the prepared muffin holes and bake for 20 minutes or until golden and cooked through. Set aside for 5 minutes to cool slightly, then transfer to a wire rack to cool.

TIP These muffins can be frozen for up to 1 month, wrapped well in plastic wrap. Thaw at room temperature.

Chia, date & walnut slice

PREPARATION TIME: 10 MINUTES (+ CHILLING)
COOKING TIME: NIL
MAKES 16 WEDGES

150 g (5½ oz/1½ cups) rolled (porridge) oats
90 g (3¼ oz/¾ cup) walnut halves
35 g (1¼ oz/½ cup) shredded coconut
45 g (1¾ oz/¼ cup) white or black chia seeds
1 teaspoon ground cinnamon
1 tablespoon unsweetened cocoa powder, sifted
2 teaspoons natural vanilla extract
425 g (15 oz) pitted dried dates, chopped
Pure icing (confectioners') sugar, to serve

* *This slice is easy to make, wheat free and satisfies sweet cravings perfectly, making it a perfect snack or after-dinner treat with coffee.*

1 Put the oats, walnuts, coconut, chia seeds, cinnamon, cocoa and vanilla in a food processor and process until all the ingredients are finely chopped.

2 With the motor running, start adding the dates a few pieces at a time, processing until all the dates have been added and the mixture is starting to come together. Use clean hands to bring it together completely, adding 1–2 teaspoons of cold water if necessary.

3 Press the mixture evenly into a round 20 cm (8 inch) cake tin, smoothing the surface firmly with the back of a metal spoon. Cover and refrigerate until ready to serve.

4 To serve, dust with icing sugar and cut into thin wedges.

TIP This slice will keep, covered with plastic wrap, in the refrigerator for up to 2 weeks.

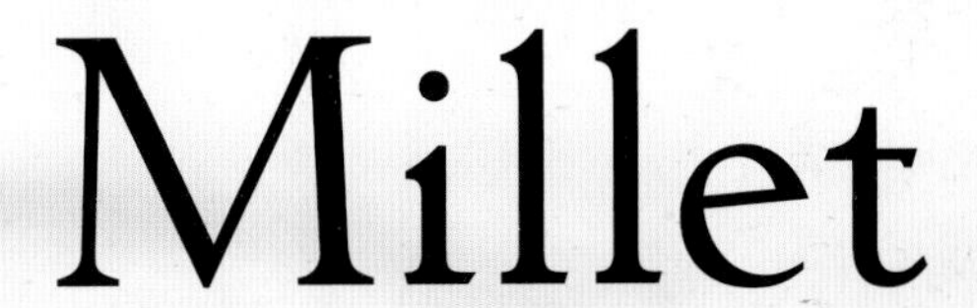

Millet

NUTRITIONAL INFORMATION
(per 175 g/6 oz/1 cup cooked millet)

Energy: 867 kilojoules, 207 calories **Protein:** 6.1 g **Total fat:** 1.7 g

Saturated fat: 0.3 g **Carbohydrate:** 41.2 g **Sugars:** 0.2 g

Dietary fibre: 2.3 g **Cholesterol:** 0 mg **Iron:** 1 mg **Phosphorus:** 174 mg

Magnesium: 77 mg **Manganese:** 0.5 mg **Sodium:** 3.5 mg

Millets are a group of small, round seeded cereal grains thought to have originated in western Africa. They have been important food staples throughout human history and have been cultivated in East Asia for the last 10,000 years. In fact, these ancient grains pre-date wheat and rice as a cultivated grain.

Because they can grow in harsh environments where other grain crops do not grow well, millets are a major food source in arid and semi-arid regions of the world, including India and Africa, where millet flour, sometimes mixed with sorghum flour, is made into a flatbread (roti). In 2009, some 26 million tons of millet were produced around the world, with India and Nigeria being the largest producers.

Millet also features in the traditional cuisine and cultures of many other countries — millet porridge, sweetened with milk and sugar is a traditional food in Russia; candied millet puffs are a speciality of Osaka, Japan; 'tongba' is a millet-based alcoholic brew in Nepal; millet is boiled with apples and honey in Germany; and in China millet is made into soup or mixed with brown sugar and used by nursing mothers to aid in milk production.

Nutritionally, millet is gluten free and suitable for those with coeliac disease or other forms of allergies/intolerances to wheat. It is also easy to digest and was traditionally used to feed the sick, but as it is a mild thyroid peroxidase inhibitor it should not be consumed in great quantities by those with thyroid disease.

Millet is a good source of manganese and contains a moderate amount of dietary fibre. It also contains the minerals phosphorus, zinc, copper and various phytochemicals including lignans, phenolic acids, phytic acid, plant sterols and saponins.

Millet is available as millet grain (hulled), millet meal, puffed millet and millet flour. The grains are tiny and are similar in appearance to yellow mustard seeds. They have a mild, buttery, slightly corn-like taste, and toasting the grain before cooking helps bring out this flavour.

Millet works beautifully with ingredients such as lime, chilli, herbs and spices. It can be used in salads instead of rice, quinoa or couscous, in pilafs, as a breakfast cereal, added to soups and stews and even to make a polenta-like dish. Once cooked it does retain quite an *al dente* texture, and the grains do separate on standing.

Millet meal can be used in baked goods as you would use polenta, and it has a similar texture. It is delicious combined with almond meal to make a gluten-free cake.

Puffed millet is ideal to add to home-made muesli or baked goods such as muesli bars.

Millet flour can be used in flat breads that do not require leavening, or it can replace some of the wheat flour in cakes and cookies. It gives a distinct golden colour and buttery taste to baked goods.

To cook millet

Heat a large saucepan over medium–high heat, add 210 g (7½ oz/1 cup) hulled millet and cook, stirring, for 3 minutes or until fragrant. Add 500 ml (17 fl oz/2 cups) water and bring to the boil. Reduce the heat to low, cover and simmer for 20 minutes or until the water is absorbed. Remove from the heat and set aside, still covered, to steam for 10 minutes. Fluff the grains with a fork and transfer to a large bowl to cool completely.

NOTE Sometimes you may find you need to add an extra 125 ml (4 fl oz/½ cup) water during cooking.

gluten-free

Millet salad with chilli lime prawns

PREPARATION TIME: 20 MINUTES (+ COOLING)
COOKING TIME: 40 MINUTES
SERVES 4

1/2 teaspoon dried red chilli flakes
2 teaspoons finely grated lime zest
2 tablespoons olive oil
20 raw large prawns (shrimp), peeled and deveined with tails left intact
210 g (7 1/2 oz/1 cup) hulled millet
2 large corn cobs
1 bunch asparagus, trimmed and halved
1 firm ripe avocado, peeled, stone removed and diced
1/2 cup coriander (cilantro) leaves, coarsely chopped
Lime wedges, to serve

DRESSING

2 tablespoons lime juice
2 tablespoons olive oil
1 long red chilli, seeded and finely chopped
1/2 teaspoon caster (superfine) sugar

1 Put the chilli flakes, lime zest and 1 tablespoon of the olive oil in a shallow glass or ceramic bowl. Add the prawns and stir to coat, then cover and set aside.

2 Heat a large saucepan over medium–high heat, add the millet and cook, stirring, for 3 minutes or until fragrant.

3 Add 500 ml (17 fl oz/2 cups) water to the millet and bring to the boil. Reduce the heat to low, cover and simmer for 15–20 minutes or until the water is absorbed. Remove from the heat and set aside, covered, to steam for 10 minutes. Fluff the grains with a fork and transfer to a large bowl to cool completely.

4 Heat a large chargrill pan over high heat. Drizzle the corn and asparagus with the remaining oil. Cook the corn, turning, for 6–8 minutes or until lightly charred and tender. Remove and set aside to cool slightly, then cut the kernels from the cobs. Cook the asparagus and prawns for 2 minutes each side or until the asparagus is tender-crisp and the prawns are just cooked through. Set the prawns aside. Add the asparagus, corn kernels, avocado and coriander to the millet.

5 To make the dressing, put all the ingredients in a small bowl and whisk until well combined and the sugar is dissolved. Add to the salad and toss gently to combine. Season with sea salt and freshly ground black pepper, to taste.

6 To serve, divide the salad among serving plates, top with the chargrilled prawns and serve with lime wedges.

* *Buttery and almost corn-like, millet is a good match with Mexican flavours such as lime, chilli and coriander.*

Millet is relatively quick to cook, gluten free and has a lovely al dente *texture once cooked.*

Creamy parmesan millet with ratatouille

PREPARATION TIME: 20 MINUTES
COOKING TIME: 50 MINUTES
SERVES 4

210 g (7½ oz/1 cup) hulled millet

500 ml (17 fl oz/2 cups) chicken or vegetable stock (see tips)

70 g (2½ oz/⅔ cup) finely grated parmesan cheese, plus coarsely grated parmesan, to serve

RATATOUILLE

2 tablespoons olive oil

1 red onion, diced

2 garlic cloves, crushed

1 red capsicum (pepper), seeded and cut into 2 cm (¾ inch) dice

2 medium zucchini (courgettes), trimmed and cut into 2 cm (¾ inch) dice

1 small eggplant (aubergine), trimmed and cut into 2 cm (¾ inch) dice

400 g (14 oz) tin chopped tomatoes

2 teaspoons thyme leaves

** When I first heard millet could be used to make a soft polenta-like dish, I was sceptical. However, I could not believe the end result — creamy, rich and best of all, unlike polenta it does not become stiff on standing. The trick is to cook it slowly with a higher ratio of liquid than usual.*

1 Heat a large saucepan over medium–high heat, add the millet and cook, stirring, for 3 minutes or until fragrant.

2 Add the stock and 500 ml (17 fl oz/2 cups) water to the millet and bring to the boil over high heat. Reduce the heat to low, cover and simmer, stirring occasionally, for 45 minutes or until thick and creamy. Add a little extra water during cooking if it looks too dry. Stir through the parmesan and season with sea salt and freshly ground black pepper.

3 Meanwhile, to make the ratatouille, heat the olive oil in a large pan over medium heat. Add the onion and cook, stirring occasionally, for 5 minutes or until softened. Add the garlic and cook, stirring, for 1 minute more.

4 Increase the heat to high, add the capsicum, zucchini and eggplant and cook, stirring occasionally, for 5 minutes or until the vegetables are golden. Add the tomatoes and thyme, reduce the heat to low and simmer for 10 minutes or until the vegetables are tender, adding a little water if the sauce is becoming too thick. Season with sea salt and freshly ground black pepper, to taste.

5 To serve, divide the parmesan millet among 4 shallow bowls, top with some ratatouille and sprinkle with the coarsely grated parmesan.

TIPS If you want this dish to be gluten free, make sure you use gluten-free stock.

You can also make millet porridge using the same method. Simply toast the millet as in step 1 and then follow step 2, replacing the stock with water. Omit the parmesan and serve with a drizzle of honey and a little milk.

Millet-stuffed roast chicken

PREPARATION TIME: 20 MINUTES (+ COOLING)
COOKING TIME: 1 HOUR 20 MINUTES
SERVES 4

2 tablespoons olive oil
1 small brown onion, finely chopped
125 g (4½ oz) bacon rashers (see tips), rind removed, diced
2 garlic cloves, crushed
175 g (6 oz/1 cup) cooked millet (see page 119)
2 tablespoons coarsely chopped parsley
1 tablespoon finely chopped sage
1 teaspoon finely grated lemon zest
1.6 kg (3 lb 8 oz) free-range chicken
Roast potatoes and steamed greens, to serve

✱ Millet works wonderfully as a stuffing, as it absorbs the flavours and juices from the chicken. It is also a gluten-free alternative to a traditional breadcrumb stuffing. Using millet in a stuffing is a great way to introduce it to people, as they won't know it's in there but will enjoy the flavour.

1 Heat 1 tablespoon of the olive oil in a large frying pan over medium–high heat. Add the onion and bacon and cook, stirring occasionally, for 5 minutes or until golden. Add the garlic and cook, stirring, for 1 minute more. Transfer the mixture to a large bowl and add the millet, herbs and lemon zest. Season with sea salt and freshly ground black pepper, then set aside to cool.

2 Preheat the oven to 220°C (425°F/Gas 7). Rinse the chicken and pat dry with paper towel. Spoon the stuffing into the cavity of the chicken. Tie the legs together with kitchen string.

3 Place the chicken, breast side up, in a large roasting pan. Drizzle with the remaining olive oil and season with salt and pepper. Roast for 20 minutes, then reduce the oven temperature to 200°C (400°F/Gas 6) and cook for a further 50 minutes or until the juices run clear when you prick the thickest part of the thigh. (Cover the chicken with foil if it is browning too quickly.)

4 Remove the chicken from the oven and set aside for 10 minutes to rest before carving. Serve with roast potatoes and steamed greens.

TIPS If you want this dish to be gluten free, make sure you use gluten-free bacon.

I think millet works particularly well with poultry or pork, so you could also try this stuffing in spatchcock, quail, turkey or a rolled pork loin.

Millet, broad beans & peas with marinated feta

PREPARATION TIME: 15 MINUTES (+ COOLING)
COOKING TIME: 30 MINUTES
SERVES 4

210 g (7½ oz/1 cup) hulled millet
500 g (1 lb 2 oz) fresh or frozen broad (fava) beans
155 g (5½ oz/1 cup) podded fresh green peas
75 g (2¾ oz) snow pea (mangetout) shoots
2 tablespoons snipped chives
100 g (3½ oz) marinated feta cheese, drained and crumbled
2 preserved lemon quarters, flesh and white pith removed, thinly sliced

DRESSING
2 tablespoons olive oil (see tip)
2 tablespoons lemon juice
Pinch of caster (superfine) sugar

** This salad makes a great accompaniment to grilled (broiled) meats or seafood. To make it a complete (balanced) meal, flake a hot-smoked salmon fillet through the salad or add a handful of nuts.*

Toasting the millet before cooking really brings out its buttery, corn-like flavour.

1 Heat a large saucepan over medium–high heat, add the millet and cook, stirring, for 3 minutes or until fragrant.

2 Add 500 ml (17 fl oz/2 cups) water to the millet and bring to the boil. Reduce the heat to low, cover and simmer for 15–20 minutes or until the water is absorbed. Remove from the heat and set aside, covered, for 10 minutes to steam. Fluff the grains with a fork and transfer to a large bowl to cool completely.

3 Blanch the broad beans and peas in a saucepan of boiling water for 2–3 minutes or until tender-crisp. Refresh under cold running water, then drain well. Peel away and discard the outer skins from the broad beans.

4 Add the peas, broad beans, snow pea shoots, chives and half the feta to the millet and stir to combine. To make the dressing, put the olive oil, lemon juice and sugar in a small bowl and whisk to combine.

5 Add the dressing to the salad and toss to combine. Season with sea salt and freshly ground black pepper, to taste. Serve topped with the remaining feta and sprinkled with the preserved lemon.

TIPS Try using some of the oil from the feta jar in the dressing — it is usually olive oil combined with garlic and herbs and will add extra flavour to your salad.

Any leftovers can be kept in an airtight container in the refrigerator for lunch the following day.

MILLET, SPINACH &
CASHEW PILAF WITH LAMB
CUTLETS (PAGE 130)

Millet, spinach & cashew pilaf with lamb cutlets

PREPARATION TIME: 20 MINUTES
COOKING TIME: 35 MINUTES
SERVES 4

210 g (7½ oz/1 cup) hulled millet

1 tablespoon olive oil, plus extra, to brush

1 brown onion, finely chopped

2 garlic cloves, crushed

1 tablespoon korma curry paste (see tips)

500 ml (17 fl oz/2 cups) chicken stock (see tips)

8 lamb cutlets, french trimmed

100 g (3½ oz) baby spinach leaves

40 g (1½ oz/¼ cup) cashew nuts, lightly toasted, coarsely chopped

2 tablespoons currants

1 tablespoon lemon juice

Greek-style yoghurt, to serve

* *I love to use millet in pilafs instead of white rice, as it has so much more flavour and retains its texture. In comparison to white rice, millet has over twice as much fibre and almost twice as much protein!*

1 Heat a frying pan over medium–high heat, add the millet and cook, stirring, for 3 minutes or until fragrant. Remove from the heat.

2 Heat the olive oil in a large heavy-based saucepan over medium heat. Cook the onion, stirring occasionally, for 5 minutes or until softened. Add the garlic and curry paste and cook, stirring, for 1 minute or until fragrant.

3 Add the toasted millet and stock and bring to the boil. Reduce the heat to low, cover and simmer for 25 minutes or until the liquid is absorbed. Remove from the heat and set aside, covered, for 10 minutes to steam.

4 Meanwhile, heat a large chargrill pan over high heat. Brush the cutlets with a little extra oil and cook for 2 minutes each side or until lightly charred and cooked to your liking.

5 Stir the spinach leaves, cashews, currants and lemon juice through the pilaf and season with sea salt and freshly ground black pepper, to taste. Divide the pilaf among serving plates and top each with 2 lamb cutlets. Serve with a dollop of yoghurt.

TIPS You can use any Indian-style curry paste in this recipe.

If you want this dish to be gluten free, make sure you use gluten-free curry paste and stock.

Healing chicken & millet soup

PREPARATION TIME: 20 MINUTES
COOKING TIME: 2 HOURS 20 MINUTES
SERVES 4

1.6 kg (3 lb 8 oz) free-range chicken
1 large onion, coarsely chopped
1 large carrot, peeled and coarsely chopped
3 celery stalks, chopped
10 black peppercorns
3 flat-leaf (Italian) parsley stalks
2 teaspoons olive oil
1 leek, trimmed, white part only, halved and thinly sliced
70 g (2½ oz/⅓ cup) hulled millet
Snipped chives, to garnish

* *This soup is both soothing and nurturing, the perfect food for when you are unwell. Millet is historically known for its therapeutic uses, as it is gluten free and very easily digested.*

1 Rinse the cavity of the chicken and place in a large saucepan with the onion, carrot, celery, peppercorns and parsley. Add enough cold water to just cover the chicken and slowly bring to the boil over high heat. Reduce the heat to low, skim any scum and oil that comes to the surface and simmer for 1 hour. Carefully remove the chicken and set aside to cool slightly. Return the stock to the heat and simmer for a further 30 minutes.

2 Strain the stock, discarding the solids, and set aside. When the chicken is cool enough to handle, remove the skin and bones and shred the breast meat (see tips). Set aside.

3 Place a large clean saucepan over medium heat and add the olive oil and leek. Cook, stirring occasionally, for 5–6 minutes or until softened. Add 1.5 litres (52 fl oz/6 cups) of the reserved stock (see tips). Add the millet and simmer for 30 minutes or until tender. Add the shredded chicken and simmer for a further 5 minutes. Season with sea salt and freshly ground black pepper, to taste. Serve garnished with chives.

TIPS The rest of the chicken meat can be shredded and used to make chicken sandwiches or added to salads.

Any remaining stock can be frozen and used to make risotto, sauces or soup.

This soup is suitable to freeze.

Orange & almond syrup cake

PREPARATION TIME: 30 MINUTES (+ COOLING)
COOKING TIME: 45–50 MINUTES
SERVES 10–12

185 g (6½ oz) unsalted butter, at room temperature
275 g (9¾ oz/1¼ cups) caster (superfine) sugar
4 eggs
130 g (4½ oz/1 cup) millet meal
3 teaspoons baking powder (see tips)
250 g (9 oz/2½ cups) almond meal
2 teaspoons finely grated orange zest
Greek-style yoghurt, to serve

SYRUP

1 orange, plus 125 ml (4 fl oz/½ cup) freshly squeezed orange juice
110 g (3¾ oz/½ cup) caster (superfine) sugar

** I like to use millet meal in cakes where I would typically use polenta. Polenta and millet are both gluten-free, but I prefer the texture of millet meal because it is not as coarse.*

1 Preheat the oven to 180°C (350°F/Gas 4). Lightly grease a round 22 cm (8½ inch) springform cake tin and line the base and side with non-stick baking paper.

2 Use an electric mixer to beat the butter and sugar until pale and creamy. Add the eggs one at a time, beating well after each addition. Fold in the millet meal, baking powder, almond meal and orange zest until well combined.

3 Spoon the mixture into the prepared tin and smooth the surface with the back of the spoon. Bake for 45–50 minutes, covering the top of the cake with foil after 30 minutes, or until a skewer inserted into the centre comes out clean. Remove and set aside to cool for 10 minutes, then carefully remove the side of the tin and transfer to a wire rack over a metal tray.

4 Meanwhile, to make the syrup, finely grate the zest of the orange (reserve the orange). Put the zest, juice and sugar in a small saucepan and cook, stirring, over low heat until the sugar has dissolved. Increase the heat to high and bring to the boil, then reduce the heat and simmer for 2 minutes.

5 Remove the syrup from the heat and carefully brush half the syrup over the warm cake, then set aside to cool completely. Brush the cake with the remaining syrup. Use a small sharp knife to remove the white pith from the orange and cut the flesh into segments. Serve the cake with the yoghurt and orange segments.

TIPS If you want this cake to be gluten free, make sure you use gluten-free baking powder.

Try adding a few drops of orange blossom water to the orange syrup for a Middle Eastern flavour.

To make small individual cakes, line a 12-hole 80 ml (2½ fl oz/⅓ cup) muffin tin with paper cases. Divide the mixture among the lined holes and bake for 20 minutes or until cooked through.

Crystallised ginger millet biscuits

PREPARATION TIME: 25 MINUTES
COOKING TIME: 15 MINUTES
MAKES ABOUT 24

110 g (3¾ oz) unsalted butter, at room temperature

110 g (3¾ oz/½ cup, firmly packed) light brown sugar

115 g (4 oz/⅓ cup) golden syrup

135 g (4¾ oz/1 cup) millet flour

110 g (3¾ oz/¾ cup) plain (all-purpose) flour

1 tablespoon ground ginger

1 teaspoon ground cinnamon

75 g (2¾ oz/⅓ cup) crystallised ginger, finely chopped, plus 55 g (2 oz/¼ cup) extra, to garnish

* *These biscuits remind me of old-fashioned gingernut biscuits, but they are so much better! They are crisp on the outside, but still a little chewy in the centre. The millet flour gives a buttery flavour and adds to their golden colour.*

1 Preheat the oven to 180°C (350°F/Gas 4). Line 2 large baking trays with non-stick baking paper.

2 Use an electric mixer to beat the butter, sugar and golden syrup in a large bowl until pale and creamy. Sift the millet flour, plain flour, ground ginger and cinnamon together, add to the butter mixture with the crystallised ginger and stir until well combined.

3 Roll tablespoons of the mixture into balls and place on the lined trays, leaving 4 cm (1½ inches) between each to allow for spreading. Flatten the balls slightly with clean fingers, then garnish each with a little extra crystallised ginger. Bake for 12–15 minutes, turning the trays halfway through cooking, or until the biscuits are golden.

4 Cool the biscuits on the trays for 10 minutes before transferring to a wire rack to cool completely.

TIP Keep in an airtight container for up to 1 week.

Oats

NUTRITIONAL INFORMATION
(per 235 g/8½ oz/1 cup rolled oats cooked in water)

Energy: 695 kilojoules, 166 calories Protein: 5.9 g

Total fat: 3.6 g Saturated fat: 0.7 g Carbohydrate: 31.8 g

Sugars: 0.6 g Dietary fibre: 4.0 g Cholesterol: 0 mg

Iron: 2.1 mg Magnesium: 63.2 mg Manganese: 1.4 mg

Phosphorus: 180 mg Selenium: 12.6 mg Sodium: 9.4 mg

oats

Oats draw their ancestry from the wild red oat, a plant originating in the Near East. They have been cultivated for more than 2000 years in various regions throughout the world, particularly Europe where they were a dietary staple in many countries, including Great Britain, Germany and Scandinavia. Scottish settlers introduced oats to North America in the 1600s.

Although oats are a familiar food to many people, most do not realise just how nutritious they are. Simple, economical and easy to source, oats and oat products can be used not only for porridge and muesli, but an amazing array of dishes.

After harvesting, oats are roasted, cleaned and hulled. This process does not remove the bran and germ, which are a concentrated source of fibre and nutrients.

Oats contain more soluble fibre than any other grain, including beta-glucans, a type of soluble fibre that has been proven to help lower cholesterol and therefore possibly reduce the risk of heart attack.

The soluble fibre and protein content of oats also results in slower digestion and an extended sensation of fullness. So if you eat oats for breakfast you may find you eat less food throughout the rest of the day.

Oats can also help reduce blood sugar and insulin levels, so can be of assistance in controlling diabetes. Oats contain iron, manganese, zinc, vitamin E, folate, B vitamins, phosphorus and selenium. They also contain antioxidants that are beneficial for health such as avenanthramides, phenolic acids and phytic acid.

Oats can be found in various forms, as groats (kernels), steel-cut oats, rolled (porridge) oats, quick oats, instant oats, oat bran and oat flour. Oats become thick and creamy when cooked with liquid due to their high soluble fibre content and are naturally a little sweet, making them the perfect addition to baked goods.

Oat groats are the whole oat kernels that have not been flattened in any way. They can be used in salads as you would barley or spelt, and are available from health food stores. Steel-cut oats are groats that have been sliced and they have a chewy texture, ideal for using in porridge or cereal, although they're harder to find than rolled oats.

Rolled oats are available as either stabilised or unstabilised. Stabilised rolled oats are steamed and then rolled, and have a flatter shape than steel-cut oats. Their soft texture makes them ideal for muesli, baking, porridge and as a coating for meat or fish. Unstabilised rolled oats are not steamed before they are rolled, so they are slightly more nutritious but have a shorter shelf life as they tend to turn rancid quickly. As a result, they are not as readily available as the stabilised version. Quick oats are steamed, finely cut and rolled, making them quicker to cook and with a softer texture than rolled oats. Instant oats are similar to quick oats but often have flavourings added.

Oat bran is the outer husk of an oat grain and is rich in fibre, vitamins and minerals. It is great to add to baked goods such as bread, muffins, cookies, smoothies and as a substitute for breadcrumbs in meatloaf or burgers. Oat flour is made from ground whole oats and can be used when making breads and other baked goods.

To cook oats

Place 95 g (3 1/4 oz/1 cup) rolled oats in a medium saucepan with 500 ml (17 fl oz/2 cups) cold water and bring to the boil. Reduce the heat to low and simmer, stirring occasionally, for 5 minutes or until thick and creamy.

NOTE If you are using unstabilised rolled oats they may take a little longer to cook.

Bircher muesli with blackberries & almonds

PREPARATION TIME: 15 MINUTES
(+ OVERNIGHT SOAKING)
COOKING TIME: NIL
SERVES 4

190 g (6¾ oz/2 cups) rolled (porridge) oats

375 ml (13 fl oz/1½ cups) apple juice

125 ml (4 fl oz/½ cup) milk

260 g (9¼ oz/1 cup) Greek-style yoghurt

1 large apple, peeled, cored and coarsely grated

80 g (2¾ oz/½ cup) raw almonds, chopped

2 tablespoons sunflower seeds

Fresh blackberries, to serve

Honey, to drizzle

** Oats make a fantastic breakfast as they are high in fibre and have a low GI, so they release energy slowly throughout the day. Bircher muesli is a great way to incorporate oats into your breakfast regime, especially in the warmer months when you may not feel like porridge.*

1 Put the oats, apple juice and milk in a medium bowl and stir to combine. Cover and refrigerate overnight.

2 The next morning, add the yoghurt, apple, three-quarters of the almonds and the sunflower seeds and stir to combine. Divide among serving bowls and top with blackberries, the remaining almonds and a drizzle of honey.

TIPS Try substituting the blackberries with fresh or thawed frozen blueberries or raspberries.

Keep, covered, in the refrigerator for up to 3 days.

Roasted sweet potato, zucchini & feta burgers

PREPARATION TIME: 20 MINUTES
(+ 20 MINUTES COOLING)
COOKING TIME: 30 MINUTES
SERVES 4

500 g (1 lb 2 oz) orange sweet potato, peeled and cut into 2 cm (3/4 inch) dice

1 teaspoon ground cumin

2 tablespoons olive oil

1 medium zucchini (courgette), trimmed and coarsely grated

50 g (1 3/4 oz/1/2 cup) rolled (porridge) oats, plus 70 g (2 1/2 oz/2/3 cup), extra

100 g (3 1/2 oz) creamy feta cheese, crumbled

40 g (1 1/2 oz/1/4 cup) pine nuts, lightly toasted

2 tablespoons snipped chives

1 tablespoon finely chopped flat-leaf (Italian) parsley

75 g (2 3/4 oz/1/3 cup) hummus, to serve

4 wholegrain bread rolls, split and toasted, to serve

Baby cos (romaine) lettuce leaves and sliced tomato, to serve

These veggie burgers are packed with flavour and the oats make a lovely crisp coating. They are an excellent serve of dietary fibre and a good source of protein and unsaturated fatty acids.

1 Preheat the oven to 200°C (400°F/Gas 6). Line a large baking tray with non-stick baking paper. Place the sweet potato on the lined tray in a single layer and sprinkle with the cumin, then drizzle with 1 tablespoon of the olive oil and season with salt and pepper. Roast for 25 minutes or until golden and tender. Transfer to a large bowl, roughly mash and set aside to cool.

2 Squeeze the excess moisture from the zucchini, then add to the sweet potato with the oats, feta, pine nuts and herbs. Season with sea salt and freshly ground black pepper. Using clean hands, mix until well combined and then shape into 4 patties. Place the extra oats on a large plate and press each patty into the oats to cover on all sides.

3 Heat the remaining olive oil in a large non-stick frying pan over medium heat. Cook the patties for 3 minutes each side or until golden and crisp.

4 To serve, spread a little hummus on the base of each roll and top with lettuce, tomato and a patty. Cover with the tops of the rolls and serve immediately.

TIP Roasting the sweet potato is worth the extra effort as it gives much more flavour than steaming it. It also reduces the moisture content, resulting in a firmer burger that should not fall apart.

Lamb koftas with tzatziki, pitta bread & cucumber salad

PREPARATION TIME: 20 MINUTES
(+ 30 MINUTES SOAKING)
COOKING TIME: 8 MINUTES
SERVES 4

500 g (1 lb 2 oz) minced (ground) lamb
35 g (1¼ oz/¼ cup) oat bran
1 egg
1 small red onion, grated
40 g (1½ oz/¼ cup) pine nuts, lightly toasted and chopped
½ teaspoon dried mint
¼ teaspoon dried red chilli flakes
1 teaspoon ground cumin
1 tablespoon olive oil
Warmed pitta bread and tzatziki, to serve

CUCUMBER SALAD

2 Lebanese (short) cucumbers, peeled, halved lengthways, seeded and sliced
6 small radishes, trimmed and thinly sliced
¼ cup coarsely chopped flat-leaf (Italian) parsley
1 tablespoon olive oil
1 tablespoon lemon juice

* *Oat bran is a good substitute for breadcrumbs and it has the added bonus of being higher in dietary fibre. Look for it in the cereal section of the supermarket.*

1 Soak 12 small wooden skewers in cold water for 30 minutes.

2 Put the minced lamb, oat bran, egg, onion, pine nuts, dried mint, chilli flakes and cumin in a medium bowl and use clean hands to mix until well combined. Season with sea salt and freshly ground black pepper. Divide the mixture into 12 portions, then use slightly wet hands to shape each into a sausage shape around a soaked skewer. Set aside.

3 To make the cucumber salad, put the cucumber, radishes and parsley in a large bowl. Drizzle with the olive oil and lemon juice, toss to combine and season with salt and pepper, to taste.

4 Preheat a large chargrill pan or barbecue over high heat. Drizzle the koftas with the olive oil and cook, turning occasionally, for 6–8 minutes or until golden and cooked through.

5 Divide the koftas and cucumber salad among 4 plates and serve with warmed pitta bread and tzatziki.

TIPS You could replace the lamb with minced beef or even pork. To make this a wheat-free meal, simply omit the pitta bread.

Pork, veal, prune & prosciutto terrine

PREPARATION TIME: 20 MINUTES (+ COOLING)
COOKING TIME: 40 MINUTES
SERVES 6

1 tablespoon olive oil
1 onion, finely chopped
1 carrot, peeled and coarsely grated
1 small zucchini (courgette), trimmed and coarsely grated
2 garlic cloves, crushed
1 teaspoon ground allspice
600 g (1 lb 5 oz) minced (ground) mixed pork and veal
55 g (2 oz/½ cup) quick oats
90 g (3¼ oz/⅓ cup) chopped pitted prunes
45 g (1¾ oz/⅓ cup) pistachios
1 egg
10 thin slices prosciutto

1 Preheat the oven to 180°C (350°F/Gas 4). Spray a 7 cm (2¾ inch) deep, 10 x 20 cm (4 x 8 inch) loaf (bar) tin with olive oil.

2 Heat the olive oil in a large frying pan over medium heat. Cook the onion, carrot and zucchini, stirring occasionally, for 5–6 minutes or until softened. Add the garlic and allspice and cook, stirring, for 1 minute. Transfer to a large mixing bowl and set aside to cool.

3 Add the minced pork and veal, oats, prunes, pistachios and egg and season with salt and pepper. Using clean hands, mix the ingredients until well combined.

4 Arrange the prosciutto slices over the base and 2 long sides of the tin, allowing them to overhang the sides. Spoon the mince mixture into the prosciutto-lined tin, pressing down firmly. Fold over the overhanging prosciutto to enclose the filling.

5 Cover with foil and bake for 15 minutes, then remove the foil and bake for a further 20 minutes or until firm to touch and the juices run clear when tested with a skewer. Remove from the oven and set aside for 10 minutes, then remove from the tin and cut into thick slices to serve.

TIP Serve this terrine warm with steamed vegetables or salad for dinner, or cold with rocket (arugula) and relish for a delicious lunch.

** Quick oats can be used as a binding agent in recipes such as burgers, terrines and meatloaf, instead of breadcrumbs. I like to add some grated vegetables as well – sneaking in a few extra veggies is always a good thing, and it reduces the amount of meat required, too.*

Chicken & leek pies with crunchy oat topping

PREPARATION TIME: 30 MINUTES
COOKING TIME: 45 MINUTES
SERVES 4

1 tablespoon olive oil

1 tablespoon butter

700 g (1 lb 9 oz) skinless chicken breast fillets, cut into 2 cm (3/4 inch) dice

2 leeks, white part only, thinly sliced

1 celery stalk, trimmed and diced

2 garlic cloves, crushed

80 ml (2 1/2 fl oz/1/3 cup) white wine

2 tablespoons plain (all-purpose) flour

125 ml (4 fl oz/1/2 cup) chicken stock

80 ml (2 1/2 fl oz/1/3 cup) thin (pouring/whipping) cream

140 g (5 oz/1 cup) frozen green peas, thawed

2 tablespoons snipped chives

Green salad or steamed greens, to serve

TOPPING

75 g (2 3/4 oz/3/4 cup) rolled (porridge) oats

35 g (1 1/4 oz/1/4 cup) plain (all-purpose) flour

50 g (1 3/4 oz/1/2 cup) finely grated parmesan cheese

2 tablespoons finely chopped flat-leaf (Italian) parsley

50 g (1 3/4 oz) chilled unsalted butter, diced

1 Preheat the oven to 190°C (375°F/Gas 5). Lightly butter four 250 ml (9 fl oz/1 cup) ovenproof baking dishes.

2 Heat the oil and butter in a large, deep frying pan over high heat. Cook the chicken, in batches, for 2–3 minutes or until golden. Transfer to a plate and set aside.

3 Return the pan to medium heat and add the leeks and celery. Cook, stirring occasionally, for 5–6 minutes or until softened. Add the garlic and cook for 1 minute more. Return the chicken to the pan, add the wine and simmer until it has almost evaporated. Add the flour and cook, stirring, for 1 minute.

4 Add the stock and cream to the pan and simmer for 5 minutes or until thickened. Stir through the peas and chives and season with sea salt and freshly ground black pepper, to taste. Divide among the buttered dishes.

5 To make the topping, put the oats, flour, parmesan, parsley and a large pinch of salt in a medium bowl. Use clean fingertips to rub in the butter until the mixture resembles small clumps.

6 Sprinkle the topping evenly over the pies. Bake for 20–25 minutes or until golden. Remove and set aside for 5 minutes to cool slightly before serving with a green salad or steamed greens.

TIP You could make one large pie if you wanted to — use a 1 litre (35 fl oz/4 cup) ovenproof pie dish and bake it for 25–30 minutes.

Lemon, herb & oat crumbed fish

PREPARATION TIME: 20 MINUTES
(+ 30 MINUTES CHILLING)
COOKING TIME: 6 MINUTES
SERVES 4

145 g (5¼ oz/1½ cups) rolled (porridge) oats

2 tablespoons finely chopped flat-leaf (Italian) parsley

2 tablespoons finely chopped dill

2 teaspoons finely grated lemon zest

75 g (2¾ oz/½ cup) plain (all-purpose) flour

2 eggs, lightly whisked

4 x 150 g (5½ oz) firm white fish fillets (such as flathead or blue eye)

2 tablespoons olive oil

Mixed salad and lemon wedges, to serve

** Oats are a fabulous alternative to breadcrumbs for a healthy crumbed fish. They stick well and turn an enticing golden brown when pan-fried.*

1 Combine the oats, herbs and lemon zest in a shallow bowl. Place the flour in a second shallow bowl and the egg in a third.

2 Dust a piece of fish in the flour, then dip in the egg and finally press into the oat mixture until evenly coated. Place on a tray and repeat with the remaining fish. Cover and refrigerate for 30 minutes.

3 Heat the oil in a large non-stick frying pan over medium–high heat (see tip). Cook the fish for 2–3 minutes each side or until golden and cooked through, adding a little more oil to the pan if necessary. Serve the fish with a mixed salad and lemon wedges.

TIP Make sure the heat isn't too high or the oats may burn.

RHUBARB, HONEY & OAT MUFFINS (LEFT, PAGE 152) AND OAT, PARMESAN & THYME BISCUITS (PAGE 153)

Rhubarb, honey & oat muffins

PREPARATION TIME: 20 MINUTES
COOKING TIME: 20 MINUTES
MAKES 12

225 g (8 oz/1½ cups) self-raising flour

1 teaspoon ground cinnamon

160 g (5½ oz/1½ cups) quick oats, plus 2 tablespoons, extra

75 g (2¾ oz/⅓ cup, firmly packed) light brown sugar, plus 2 teaspoons, extra

1 egg

115 g (4 oz/⅓ cup) honey

1 teaspoon natural vanilla extract

100 g (3½ oz) unsalted butter, melted and cooled

250 ml (9 fl oz/1 cup) buttermilk

110 g (3¾ oz) thinly sliced rhubarb

** Quick oats are smaller and have a softer texture than regular oats, resulting in lighter muffins.*

1 Preheat the oven to 180°C (350°F/Gas 4). Line a 12-hole 80 ml (2½ fl oz/⅓ cup) muffin tin with paper cases.

2 Sift the flour and cinnamon into a large bowl. Stir through the oats and sugar. Whisk the egg, honey, vanilla, butter and buttermilk together in a large jug, add to the dry ingredients and stir until just combined. Stir through the rhubarb (do not over-mix). Divide the mixture among the lined muffin holes.

3 Combine the extra oats and brown sugar, and sprinkle over the tops of the muffins. Bake for 20 minutes or until golden brown. Set aside for 5 minutes to cool slightly, then transfer to a wire rack to cool completely.

TIP These muffins freeze well, so make an extra batch and keep some in the freezer for a delicious breakfast on the run or to pack in lunch boxes.

Oat, parmesan & thyme biscuits

PREPARATION TIME: 25 MINUTES
(+ 30 MINUTES CHILLING)
COOKING TIME: 25 MINUTES
MAKES ABOUT 30

130 g (4½ oz/1 cup) oat bran

150 g (5½ oz/1 cup) plain (all-purpose) flour

¼ teaspoon baking powder

2 tablespoons light brown sugar

1 tablespoon chopped thyme leaves

70 g (2½ oz/⅔ cup) finely grated parmesan cheese, plus 25 g (1 oz/¼ cup), extra

100 g (3½ oz) chilled unsalted butter, diced

1 egg yolk

60 ml (2 fl oz/¼ cup) iced water

** Oat bran is the outer husk of the oat grain and contains most of the grain's dietary fibre. Add it to biscuits or muffins, sprinkle it over cereal or add it to smoothies for a simple way to increase fibre in your diet. Oat bran can be found in the cereal section of the supermarket.*

1 Preheat the oven to 180°C (350°F/Gas 4). Line 2 large baking trays with non-stick baking paper.

2 Put the oat bran, flour, baking powder, a large pinch of salt and the sugar in a food processor and pulse to combine. Add the thyme, parmesan and butter and process until the mixture resembles fine breadcrumbs. Add the egg yolk and water and process until the mixture comes together.

3 Divide the dough into 2 equal portions and shape each into a flat disc. Cover with plastic wrap and refrigerate for 30 minutes.

4 Use a lightly floured rolling pin to roll out a portion of dough on a lightly floured work surface until 5 mm (¼ inch) thick. Use a round 5 cm (2 inch) cutter to cut the dough into discs, re-rolling and cutting the scraps to make more discs, and place on the lined tray, 3 cm (1¼ inches) apart. Sprinkle with a little extra parmesan and bake for 12 minutes or until light golden. Remove and set aside to cool completely on the trays. Repeat with the remaining dough.

TIPS These biscuits make a great addition to a cheese platter. Try them with quince paste and blue cheese or vintage cheddar. Keep in an airtight container for up to 1 week.

Pear, strawberry & ginger crumble

PREPARATION TIME: 20 MINUTES (+ 5–10 MINUTES COOLING)
COOKING TIME: 25–30 MINUTES
SERVES 6

1 kg (2 lb 4 oz) firm ripe pears, peeled, cored and quartered
1 tablespoon lemon juice
250 g (9 oz) strawberries, hulled and halved
2 tablespoons caster (superfine) sugar
1 teaspoon finely grated ginger
1 tablespoon plain (all-purpose) flour
Vanilla ice cream or cream, to serve

CRUMBLE

50 g (1¾ oz/½ cup) rolled (porridge) oats
35 g (1¼ oz/½ cup) shredded coconut
50 g (1¾ oz/⅓ cup) plain (all-purpose) flour
1 teaspoon ground ginger
75 g (2¾ oz/⅓ cup, firmly packed) light brown sugar
80 g (2¾ oz) chilled butter, diced

* *This crumble is quick and easy to make, as the filling does not require pre-cooking. If your pears are not ripe, they make take a little longer to cook. Simply cover the dish with foil if the topping starts becoming too golden during the extra cooking time.*

1 Preheat the oven to 180°C (350°F/Gas 4). Lightly butter six 250 ml (9 fl oz/1 cup) ovenproof dishes.

2 Thinly slice the pears, place in a large mixing bowl with the lemon juice and toss to combine. Add the strawberries, sugar, ginger and flour and toss to coat. Spoon the filling into the buttered dishes.

3 To make the crumble, put the oats, coconut, flour, ginger and sugar in a large mixing bowl. Use clean fingertips to rub in the butter until the mixture resembles small clumps. Sprinkle the crumble mixture evenly over the filling in each dish.

4 Cover the dishes with foil and bake for 15 minutes, then remove the foil and bake for a further 10–15 minutes or until the topping is golden and the pears are tender. Remove from the oven and set aside for 5–10 minutes to cool before serving with ice cream or cream.

TIPS You can substitute the rolled oats with rolled barley or spelt.

To make one large crumble, use a 1.5 litre (52 fl oz/6 cup) ovenproof baking dish and bake for 20–25 minutes, covering the dish with foil if the topping is browning too quickly.

Cranachan with raspberries & honey

PREPARATION TIME: 15 MINUTES (+ COOLING)
COOKING TIME: 5 MINUTES
SERVES 4

95 g (3¼ oz/1 cup) rolled (porridge) oats

2 tablespoons light brown sugar

200 ml (7 fl oz) thickened (whipping) cream

260 g (9¼ oz/1 cup) Greek-style yoghurt

90 g (3¼ oz/¼ cup) honey, plus extra, to drizzle (optional)

250 g (9 oz/2 cups) fresh raspberries

1–2 tablespoons whisky

** Cranachan is a traditional Scottish dessert of toasted oats, cream, whisky and often raspberries. I like to use a combination of cream and thick yoghurt for a slightly less rich version. Toasting the oats first brings out their delicious nutty flavour.*

1 Preheat the grill (broiler) on high. Spread the oats over a baking tray in a single layer and sprinkle evenly with the sugar. Grill for 3 minutes, stirring occasionally, or until just golden (watch carefully to ensure they do not burn). Remove and set aside to cool.

2 Use an electric mixer with a whisk attachment to whisk the cream until soft peaks form. Fold through the yoghurt, honey, half the raspberries and three-quarters of the cooled oats. Add the whisky, to taste, and gently fold in.

3 Spoon the mixture into four small, about 185 ml (6 fl oz/¾ cup) serving glasses, and top with the remaining raspberries and oats. Drizzle with a little extra honey, if desired.

Spelt & Kamut

NUTRITIONAL INFORMATION
(per 195 g/6¾ oz/1 cup cooked spelt)

Energy: 1030 kilojoules, 246 calories **Protein:** 10.7 g **Total fat:** 1.6 g

Saturated fat: 0 g **Carbohydrate:** 51.3 g **Sugars:** 0 g **Dietary fibre:** 7.6 g

Cholesterol: 0 mg **Niacin:** 5.0 mg **Iron:** 3.2 mg **Thiamin:** 0.2 mg

Magnesium: 95.1 mg **Manganese:** 2.1 mg **Phosphorus:** 291 mg **Sodium:** 9.7 mg

NUTRITIONAL INFORMATION
(per 160 g/5½ oz/1 cup kamut flour)

Energy: 2345 kilojoules, 558 calories **Protein:** 22.6 g **Total fat:** 4.8 g

Saturated fat: 0.96 g **Carbohydrate:** 94.4 g **Sugars:** 2.6 g **Sodium:** 12.3 mg

Accredited and extensive nutritional analysis for kamut is limited due to its small production and relatively recent availability.

Spelt

Spelt (*Triticum spelta*) is an ancient grain with a long and complex history dating back thousands of years in both Europe and the Middle East. It was one of the first grains used to make bread, and is mentioned in the Old Testament as well as various Roman texts. It was a staple grain with symbolic importance in ancient civilisations such as Greece and Rome.

As populations migrated, spelt became increasingly popular in Europe up until the end of the 1800s. However, when new developments in farming practices allowed common wheat to be harvested in a single process, spelt production dropped off. Its tough outer husk still required further processing and this made it less popular. It wasn't until the 1980s in Europe that spelt finally experienced a rebirth, and it began being farmed in Australia in 1988.

Spelt is rich in vitamins B_2 and B_3, dietary fibre, manganese, phosphorus, niacin, thiamin and copper. Spelt also contains more protein, fats and crude fibre than common wheat.

Kamut

Kamut is the trademarked name for an ancient heirloom grain whose original name is khorasan. The exact history of khorasan is a little unclear, but we do know it originated in areas around Egypt. Elaborate stories exist about its origins, such as its presence in the tombs of Egyptian pharaohs, but they are a little dubious!

Kamut is a large grain that is higher in protein than common wheat, as well as several vitamins and minerals including selenium, magnesium and zinc.

Today, consumption of spelt and kamut is once again growing rapidly as people are becoming aware of the amazing nutritional benefits of these ancient grains. Spelt and kamut are both relatives of the common wheat grain, but their properties are quite different. Even though they are members of the wheat family, they can often be tolerated by those with wheat sensitivity.

Spelt and kamut are whole grains, containing all three parts of the grain: the bran, germ and endosperm. This is unlike wheat, as processing usually removes the bran and germ. Spelt and kamut therefore contain a wider range of nutrients than common wheat.

Spelt is available in its hulled whole-grain form (often called berries), as rolled flakes and a flour, which can be used in many of the same ways as wheat flour. Kamut is a little harder to find than spelt, and is available in its hulled whole-grain form and as a flour.

Spelt or kamut grain can be used in salads as you would use rice, pearl barley or couscous.

Products using spelt or kamut flour such as pasta and bread are now readily available and are becoming increasingly popular. Both spelt and kamut flour can be used for bread making, and each makes a highly nutritious loaf. Spelt has a delicious nutty flavour, whereas kamut has quite a rich buttery taste. Bread products made with spelt or kamut flour require a little less water and less kneading than regular wheat flour.

Spelt or kamut flour can be substituted or added to other baked goods such as muffins, pancakes and cookies. Spelt or kamut pasta is a great alternative to wholemeal (whole-wheat) pasta and both are ideal for pasta salads, teaming beautifully with ingredients such as pine nuts, basil, olives and vine-ripened tomatoes to name a few.

To cook spelt and kamut

To prepare spelt or kamut grain, place 200 g (7 oz/1 cup) grain in a saucepan with 1.5–2 litres (52–70 fl oz/6–8 cups) water. Bring to the boil, then reduce the heat and simmer for 45–50 minutes or until the grains are *al dente*. Rinse under cold running water; drain well.

Home-made pizzas with salami, tomato, red onion & rocket

PREPARATION TIME: 25 MINUTES
(+ 10 MINUTES STANDING
AND 1 HOUR RISING)
COOKING TIME: 20–25 MINUTES
MAKES 4 PIZZAS

80 ml (2½ fl oz/⅓ cup) tomato passata (puréed tomatoes)
100 g (3½ oz) thinly sliced salami
2 large vine-ripened tomatoes, diced
1 small red onion, thinly sliced
150 g (5½ oz) bocconcini (fresh baby mozzarella cheese), thinly sliced
100 g (3½ oz) baby rocket (arugula) leaves, to serve
Balsamic vinegar, to drizzle (optional)

PIZZA DOUGH
1 teaspoon honey
450 g (1 lb/3 cups) spelt flour
250 ml (9 fl oz/1 cup) tepid water
7 g (⅛ oz/2 teaspoons) dried yeast
1½ tablespoons olive oil, plus extra, to grease

1 To make the dough, put the honey, 75 g (2¾ oz/½ cup) of the flour and the water in a small bowl and whisk to combine. Sprinkle with the yeast and set aside for 10 minutes, until bubbling and foamy.

2 Put the remaining flour and a pinch of salt in a large bowl and add the yeast mixture and olive oil. Stir with a wooden spoon to combine, then use your hands to bring the mixture together to form a ball.

3 Turn the dough out onto a lightly floured work surface and knead for 3–4 minutes or until it is smooth and elastic. Take care not to over-knead the dough. Lightly grease a large bowl, put the dough in it and spread a little extra oil over the top of the dough to prevent a crust forming. Cover with plastic wrap and leave to rise in a warm, draught-free place for 1 hour or until the dough has doubled in size.

4 Preheat the oven to 230°C (450°F/Gas 8). Brush 2 large pizza trays with oil. Punch the dough down with your fist, then divide it into 4 equal portions. Set 2 portions aside, covered with a damp tea towel (dish towel). Roll out the remaining portions of dough to make two 25–30 cm (10–12 inch) circles. Place each one on a greased tray.

5 Spread each pizza with 1 tablespoon of tomato passata, then top each with one-quarter of the salami, tomatoes and onion.

6 Bake the pizzas for 5 minutes, top each with one-quarter of the bocconcini, then return to the oven (swapping the trays around to ensure even cooking) and cook for 5–7 minutes more or until the toppings are golden and the bases are crisp. While they are cooking you can prepare the next 2 pizzas if you like. Remove the cooked pizzas from the oven and serve topped with rocket and a drizzle of balsamic, if desired. When you are ready, cook the next 2 pizzas, adding the bocconcini after 5 minutes of cooking, as before.

** Kamut and spelt flour are both great to use in baked goods where you would typically use wheat flour, such as cakes, breads and pizza bases. They do contain gluten, however the gluten does not require as much kneading as regular wheat flour.*

TIP For a vegetarian pizza try cubes of roasted pumpkin (winter squash) with thinly sliced zucchini (courgette), crumbled fresh ricotta cheese and chopped parsley.

Warm spelt salad with roast tomato & eggplant

PREPARATION TIME: 20 MINUTES
COOKING TIME: 45 MINUTES
SERVES 4

200 g (7 oz/1 cup) spelt

1 medium eggplant (aubergine), trimmed and cut into 1.5 cm (5/8 inch) dice

1 large zucchini (courgette), trimmed and cut into 1.5 cm (5/8 inch) dice

1 teaspoon ground cumin

60 ml (2 fl oz/1/4 cup) olive oil

250 g (9 oz) cherry tomatoes, halved

1/2 cup basil leaves, torn

1/2 cup flat-leaf (Italian) parsley leaves, coarsely chopped

100 g (3 1/2 oz) creamy feta cheese, crumbled

2 tablespoons balsamic vinegar

** Spelt is ideal to use in salads where you would usually use rice, couscous or pasta. It works beautifully with Mediterranean flavours such as tomatoes, basil and balsamic vinegar.*

1 Preheat the oven to 180°C (350°F/Gas 4). Cook the spelt in a large saucepan of lightly salted boiling water for 45 minutes or until *al dente*. Rinse under cold running water, then drain well.

2 Meanwhile, put the eggplant and zucchini on a large baking tray lined with non-stick baking paper in a single layer. Sprinkle with the cumin, drizzle with 1 tablespoon of the olive oil and season with sea salt and freshly ground black pepper. Roast for 12 minutes, then add the tomatoes to the tray, cut side up. Roast the vegetables for a further 8–10 minutes or until the eggplant is tender and the tomatoes are just wilted.

3 Put the spelt, roasted vegetables, basil, parsley and feta in a large bowl and season with salt and pepper.

4 Combine the remaining 2 tablespoons of the olive oil and the balsamic vinegar, add to the salad and gently toss to combine.

TIP You could use kamut, pearl barley, brown rice, buckwheat or quinoa instead of the spelt, though the cooking times will vary.

Kamut pasta with honey-roasted pumpkin, ricotta & crisp sage

PREPARATION TIME: 20 MINUTES
COOKING TIME: 25 MINUTES
SERVES 4

1 tablespoon honey

2 tablespoons olive oil

1/2 teaspoon dried red chilli flakes

800 g (1 lb 12 oz) butternut pumpkin (squash), peeled, seeded and cut into 1.5 cm (5/8 inch) dice

400 g (14 oz) kamut pasta

1 tablespoon unsalted butter

3 garlic cloves, very thinly sliced

16 small sage leaves

230 g (8 1/4 oz/1 cup) fresh ricotta cheese, crumbled

** Kamut pasta is available from health food stores, delicatessens and gourmet supermarkets. It has a nutty taste that works well with the sweetness of the honey and pumpkin in this dish. It also holds its shape well once cooked.*

1 Preheat the oven to 200°C (400°F/Gas 6). Combine the honey, 1 tablespoon of the olive oil and the chilli flakes. Place the pumpkin on a large baking tray lined with non-stick baking paper in a single layer. Season with sea salt and freshly ground black pepper, drizzle with the honey mixture and roast for 20–25 minutes or until it is golden and tender.

2 Meanwhile, cook the pasta in a large saucepan of lightly salted boiling water according to the packet instructions or until *al dente*. Drain well and return to the pan.

3 Heat the butter and remaining olive oil in a large frying pan over medium–high heat. Add the garlic and sage leaves and cook, stirring, for 2 minutes or until the butter and garlic are light golden and the sage is crisp.

4 Add the pumpkin, sage mixture and half the ricotta to the pasta and toss to combine. Season with salt and pepper, to taste. Serve the pasta topped with the remaining ricotta.

TIP Tossing half the ricotta through the hot pasta gives it a creamy consistency, without the fat that cream would bring. If the pasta is a little dry, try adding a little of the pasta's cooking liquid.

SEEDED SPELT BREAD
(PAGE 170)

Seeded spelt bread

PREPARATION TIME: 20 MINUTES
(+ 1 HOUR 30 MINUTES RISING)
COOKING TIME: 35 MINUTES
MAKES 1 LOAF (12–14 SLICES)

1 tablespoon honey
225 g (8 oz/1$\frac{1}{2}$ cups) white spelt flour
300 ml (10$\frac{1}{2}$ fl oz) tepid water
7 g ($\frac{1}{8}$ oz/2 teaspoons) dried yeast
225 g (8 oz/1$\frac{1}{2}$ cups) wholemeal (whole-wheat) spelt flour
1 teaspoon sea salt
1 tablespoon sesame seeds
1 tablespoon linseeds
1 tablespoon poppy seeds, plus extra, to sprinkle
2 tablespoons rolled (porridge) oats, plus extra, to sprinkle
1 tablespoon olive oil, plus extra, to grease

* *Spelt and kamut flour are ideal for bread making, as they are higher in nutrients (especially protein) than regular wheat flour and they have a surprisingly light texture when baked. Their nutty, slightly sweet flavour is similar to a light rye bread.*

1 Put the honey, 75 g (2$\frac{3}{4}$ oz/$\frac{1}{2}$ cup) of the white spelt flour and the water in a small bowl and whisk to combine. Sprinkle with the yeast and set aside for 10 minutes, until the mixture is bubbling and foamy.

2 Sift the remaining white spelt flour, the wholemeal flour and the salt into a large bowl, then return the bran in the sieve to the bowl. Stir through the sesame seeds, linseeds, poppy seeds and oats. Add the yeast mixture and olive oil. Stir with a wooden spoon to combine, then use your hands to bring the mixture together to form a ball, adding a little more water if necessary.

3 Turn the dough out onto a lightly floured work surface and knead for 3–4 minutes or until smooth and elastic. Lightly grease a large bowl, put the dough in it and spread a little extra oil over the top of the dough to prevent a crust forming. Cover with plastic wrap and leave to rise in a warm, draught-free place for 1 hour or until the dough has doubled in size.

4 Preheat the oven to 180°C (350°F/Gas 4). Brush a 10 x 20 cm (4 x 8 inch) loaf (bar) tin with oil. Punch the dough down with your fist, shape into a 20 cm (8 inch) log and place in the oiled tin. Cover loosely with plastic wrap and set aside in a warm, draught-free place until doubled in size, about 20 minutes. Sprinkle the top with the extra poppy seeds and oats.

5 Bake the bread for 30–35 minutes or until it is golden brown and the loaf sounds hollow when tapped. (Cover the loaf loosely with foil if the top is browning too quickly.) Remove from the tin and set aside on a wire rack to cool.

TIPS You can use all wholemeal spelt flour instead of the combined white and wholemeal flour, but it will result in a slightly denser loaf.

Unlike regular wheat flour, it is important not to over-knead the dough as the gluten will start to break down.

Grilled olive & anchovy yoghurt flatbreads

PREPARATION TIME: 20 MINUTES
(+ 20 MINUTES RESTING)
COOKING TIME: 25 MINUTES
MAKES 6

200 g (7 oz/3/4 cup) Greek-style yoghurt

4 anchovy fillets, finely chopped

40 g (11/2 oz/1/4 cup) pitted kalamata olives, chopped

2–3 tablespoons olive oil

1 teaspoon sea salt

225 g (8 oz/11/2 cups) spelt flour or 240 g (81/2 oz/11/2 cups) kamut flour

1/2 teaspoon baking powder

* *White spelt flour is lighter in colour and texture than wholemeal (whole-wheat) spelt flour as it is more finely milled. Either will work in this recipe.*

Although spelt and kamut are types of wheat, they are more easily digested so are suitable for some wheat-restricted diets. They are also higher in protein than wheat flour and contain slightly less kilojoules.

1 Put the yoghurt, anchovies, olives, 1 tablespoon of the oil and the salt in a large bowl and stir to combine. Sift the flour and baking powder together. Gradually add the flour and any bran in the sieve to the yoghurt mixture, stirring with a wooden spoon until the mixture starts to form a dough, then using your hands to bring it together.

2 Turn the dough out onto a lightly floured work surface and knead until smooth and elastic, about 2–3 minutes. Lightly oil a large bowl, put the dough in it and spread a little extra oil over the top of the dough to prevent a crust forming. Cover with plastic wrap and set aside for 20 minutes to rest.

3 Divide the dough into 6 equal portions. Roll each portion out on a lightly floured work surface into an oval shape, approximately 5 mm (1/4 inch) thick.

4 Heat a large chargrill pan over medium–high heat. Just before cooking, brush each flatbread with some of the remaining oil. Cook the flatbreads, one at a time, for 2 minutes each side or until lightly charred and cooked through. Remove, cover loosely with foil and keep warm while you cook the remaining flatbreads. Serve as part of a mezze plate.

TIPS You can experiment with the flavours in these flatbreads. Try replacing the anchovies or olives with chopped herbs, such as rosemary or thyme, or adding some finely grated parmesan cheese.

The best thing about flatbreads is they are quick to make and can even be cooked on the barbecue.

Little spiced fruit scrolls

PREPARATION TIME: 20 MINUTES
COOKING TIME: 30 MINUTES
MAKES 10

450 g (1 lb/3 cups) spelt flour (see tips)
1½ tablespoons baking powder
60 g (2¼ oz) chilled unsalted butter, diced
250 ml (9 fl oz/1 cup) milk
20 g (¾ oz) unsalted butter, melted
60 g (2¼ oz/¼ cup, firmly packed) light brown sugar
2 teaspoons ground cinnamon
160 g (5½ oz/1 cup) fruit medley (see tips)
1 teaspoon natural vanilla extract
2 tablespoons apricot jam, warmed

** These scrolls can be made in minutes, unlike traditional yeast bread. Their only downside is that they do not keep as well. The slightly nutty and mellow flavour of the spelt flour is delicious with spices such as cinnamon and the sweetness of dried fruit and brown sugar.*

1 Preheat the oven to 180°C (350°F/Gas 4). Brush a 20 x 30 cm (8 x 12 inch) baking tin with melted butter.

2 Sift the flour, baking powder and a large pinch of salt into a large bowl. Use your fingertips to rub in the butter until the mixture resembles coarse crumbs. Make a well in the centre, add the milk and mix with a flat-bladed knife, using a cutting action, until the mixture comes together to form a dough.

3 Turn the dough out onto a lightly floured work surface and knead briefly for 1–2 minutes or until smooth. Roll out to a rectangle about 25 x 35 cm (10 x 14 inches) and brush the surface with three-quarters of the melted butter. Combine the brown sugar and cinnamon and sprinkle evenly over the melted butter. Combine the dried fruit and vanilla and sprinkle evenly over the sugar mixture.

4 Starting at a long side, roll up the dough like a Swiss roll (jelly roll), enclosing the filling. Trim the edges and slice the dough into 10 rounds. Place, cut side up and side by side, in the greased tin and brush the top of each with the remaining melted butter. Bake for 30 minutes or until golden brown. Remove from the oven and set aside for 5 minutes, then brush the top of each scroll with warm jam. Eat while they are warm.

TIPS You can use white or wholemeal (whole-wheat) spelt flour, though wholemeal spelt flour will give a slightly denser texture.

Fruit medley is a dried fruit blend of sultanas and diced apricots, apples, peaches and pears. You can substitute the same quantity of any diced dried fruit.

These scrolls are best eaten warm on the day they are made. Reheat them in the microwave on medium (50%) for 20 seconds.

Maple & cinnamon spelt granola

PREPARATION TIME: 10 MINUTES
COOKING TIME: 40 MINUTES
MAKES ABOUT 5 CUPS

150 g (5½ oz/1½ cups) rolled spelt
150 g (5½ oz/1½ cups) rolled (porridge) oats
80 g (2¾ oz/½ cup) almonds, chopped
55 g (2 oz/⅓ cup) sunflower seeds
40 g (1½ oz/¼ cup) sesame seeds
1½ teaspoons ground cinnamon
60 ml (2 fl oz/¼ cup) pure maple syrup
2 tablespoons, firmly packed light brown sugar
2 tablespoons vegetable oil (such as sunflower or canola)
2 teaspoons natural vanilla extract
185 g (6½ oz/1 cup) mixed dried fruit (such as currants and sweetened dried cranberries)

** Rolled spelt can be used in recipes where you would typically use rolled (porridge) oats.*

1 Preheat the oven to 150°C (300°F/Gas 2). Put the rolled spelt, oats, almonds, sunflower seeds, sesame seeds and cinnamon in a large bowl and stir to combine. Transfer to a large baking tray with sides and spread out evenly.

2 Put the maple syrup, sugar and oil in a small saucepan over medium heat and cook, stirring, for 2 minutes or until the sugar has dissolved. Stir through the vanilla. Pour the hot syrup over the spelt mixture and stir until all the ingredients are evenly coated.

3 Bake the granola, stirring every 10 minutes to ensure even browning, for 30–35 minutes or until golden brown and crisp. Remove from the oven, add the dried fruit and stir to combine. Set aside to cool completely.

TIPS You could substitute rolled barley for some of the rolled oats.

It is important to stir the granola every 10 minutes or it will brown unevenly and burn. Cooking it slowly at a low temperature makes it crisp and golden without requiring too much oil or sugar.

Store in an airtight container for up to 1 month.

Chocolate & date spelt cake

PREPARATION TIME: 30 MINUTES
COOKING TIME: 45–50 MINUTES
SERVES 10–12

160 g (5½ oz) unsalted butter, softened

220 g (7¾ oz/1 cup) caster (superfine) sugar

110 g (3¾ oz/½ cup, firmly packed) light brown sugar

1 teaspoon natural vanilla extract

4 eggs, separated

115 g (4 oz/¾ cup) white or wholemeal (whole-wheat) spelt flour

2 teaspoons baking powder

30 g (1 oz/¼ cup) unsweetened cocoa powder, plus extra, to dust

160 g (5½ oz/⅔ cup) sour cream

100 g (3½ oz) dark chocolate, melted and cooled

90 g (3¼ oz) fresh pitted dates, chopped

** The nutty characteristic of spelt is a perfect match with chocolate and dates in this rich and delicious dessert cake.*

1 Preheat the oven to 180°C (350°F/Gas 4). Grease a round 24 cm (9½ inch) springform cake tin and line the base and sides with non-stick baking paper.

2 Use an electric mixer to beat the butter, sugars and vanilla until light and creamy. Add the egg yolks one at a time, beating well after each addition. Sift the spelt flour, baking powder and cocoa together, then fold into the butter mixture alternately with the sour cream. Stir through the melted chocolate and dates until just combined.

3 Use an electric mixer with a whisk attachment to whisk the egg whites in a clean, dry bowl until firm peaks form. Fold one-third of the egg whites into the chocolate mixture to loosen it, then fold in the remaining egg whites until just combined. Pour into the prepared tin and bake for 45–50 minutes or until firm to touch and there is no underlying wobble.

4 Set aside to cool slightly, then remove the side of the tin and transfer to a wire rack to cool completely. Serve dusted with extra cocoa powder.

TIPS Keep in an airtight container for up to 4 days.
Try serving this indulgent cake with Greek-style yoghurt or cream.

Buttermilk spelt pancakes with honey butter & strawberries

PREPARATION TIME: 20 MINUTES
COOKING TIME: 20 MINUTES
SERVES 4

225 g (8 oz/1½ cups) white or wholemeal (whole-wheat) spelt flour

3 teaspoons baking powder

2 tablespoons caster (superfine) sugar

2 eggs, separated

30 g (1 oz) unsalted butter, melted, plus extra, for greasing

250 ml (9 fl oz/1 cup) buttermilk

1 teaspoon natural vanilla extract

Hulled and halved strawberries, to serve

HONEY BUTTER

80 g (2¾ oz) unsalted butter, softened

2 tablespoons honey

** Spelt flour can be substituted in most recipes that contain regular wheat flour. Adding the whisked egg whites separately helps make the pancakes light and fluffy, even when using wholemeal (whole-wheat) spelt flour.*

1 To make the honey butter, put the butter and honey in a small bowl and use an electric mixer to beat until pale and fluffy. Set aside.

2 Sift the flour and baking powder into a large mixing bowl. Add the sugar and stir to combine. Make a well in the centre, add the egg yolks, melted butter, buttermilk and vanilla and whisk until smooth.

3 Use an electric mixer with a whisk attachment to whisk the egg whites in a clean, dry bowl until firm peaks form. Gently fold the egg whites into the batter until just combined.

4 Heat a large non-stick frying pan over medium heat and brush with a little extra melted butter. Pour 60 ml (2 fl oz/¼ cup) of the batter into the pan and use the back of a spoon to spread the mixture to a circle about 12 cm (4½ inches) in diameter. Repeat to make another pancake. Cook for 2 minutes or until bubbles appear on the surface of each. Turn and cook for 1–2 minutes or until cooked through.

5 Transfer the pancakes to a plate and cover loosely with foil to keep warm. Repeat with the remaining batter, brushing the pan with melted butter before cooking each batch of pancakes. Serve with strawberries and the whipped honey butter.

TIPS The honey in the honey butter can be replaced with crushed honeycomb, golden syrup or brown sugar.

For a healthier version, omit the honey butter and serve with a dollop of fresh ricotta cheese.

Barley

NUTRITIONAL INFORMATION
(per 160 g/5½ oz/1 cup cooked pearl barley)

Energy: 808 kilojoules, 193 calories **Protein:** 3.5 g **Total fat:** 0.7 g

Saturated fat: 0.1 g **Carbohydrate:** 44.3 g **Sugars:** 0.4 g

Dietary fibre: 6.0 g **Cholesterol:** 0 mg **Niacin:** 3.2 mg **Iron:** 2.1 mg

Manganese: 0.4 mg **Selenium:** 13.5 µg **Sodium:** 4.7 mg

Barley is a nutritious and versatile grain belonging to the grass family, with a history dating back more than 10,000 years to South-East Asia. During this time, barley was a staple food source for both humans and animals, and it was also used to make alcoholic beverages, as a form of currency and for medicinal purposes (barley water). Barley also played an important role in Greek and Roman culture as it was valued as a nutritious food for athletes and gladiators, and was used in bread making.

Barley looks similar to wheat, although a little paler in colour. In medieval Europe, bread made from barley and rye was peasant food, while wheat products were consumed by the upper class. Today, the health benefits of this versatile grain and its delicious nutty flavour are once again being appreciated.

Hulled barley is produced by removing the inedible fibrous outer hull. It is a nutritious and healthy food, but it can be difficult to find. Pearl barley is hulled barley with the bran removed and the grain polished. Pearl barley is less chewy and faster to cook than hulled barley, but is lower in nutrients.

Scotch or pot barley has the outer hull removed by polishing but is less polished and more nutrient dense than pearl barley and is popular in soups, hence its name.

Barley can be processed into a variety of products including flour, flakes similar to oatmeal, and grits. Sprouted barley is naturally high in maltose, a sugar that serves as the basis for both malt syrup and, when fermented, beer and other alcoholic beverages.

Barley is a great food for those with type 2 diabetes or those on a low-GI diet as it has a GI of just 25. Its high soluble-fibre content works in slowing down the absorption of sugar, lowering the overall GI value of a meal.

Barley is packed with beta-glucan soluble fibre, and is therefore linked to a reduced risk of heart disease. It also contains insoluble fibre, which helps maintain a healthy bowel and can help reduce the risk of cancers such as colon cancer.

Barley is a very good source of selenium, and a good source of phosphorus, copper and manganese. It also contains various phytochemicals including lignans, phenolic acids, phytic acid, plant sterols and saponins.

Pearl barley is by far the most common and readily available form of barley. It is delicious used in salads instead of rice, and has a slightly chewy texture and nutty taste. A winter soup is not quite the same without barley, and it also works beautifully in slow-cooked casseroles and stews. Barley can replace rice in baked puddings, but you do need to cook the barley first to reduce the overall cooking time. It also makes a great risotto that does not require constant stirring.

Barley is also available in flakes, which look similar to rolled (porridge) oats. These can be used in porridge, crumbles and cookies to name a few. Barley flour can be added to unleavened baked goods such as bread.

To cook barley

Add 200 g (7 oz/1 cup) pearl barley to 1.5 litres (52 fl oz/ 6 cups) boiling water and gently boil for 30 minutes or until *al dente*. Refresh under cold running water, then drain well. If using hulled or pot barley, increase the cooking time to 50–60 minutes.

NOTE Some people like to pre-soak barley, especially hulled or pot barley, to reduce cooking times. If you are using pearl barley this is not really necessary.

wheat-free

Poached chicken, barley, mint, zucchini & pine nut salad

PREPARATION TIME: 20 MINUTES
COOKING TIME: 45 MINUTES
SERVES 4

200 g (7 oz/1 cup) pearl barley, briefly rinsed
1 brown onion, halved
6 black peppercorns
1 teaspoon sea salt
500 g (1 lb 2 oz) skinless chicken breast fillets
2 large zucchini (courgettes), trimmed
4 baby pattypan squash, trimmed
1/2 cup mint leaves
2 tablespoons snipped chives
50 g (1 3/4 oz/1/3 cup) pine nuts, lightly toasted
2 tablespoons currants

DRESSING

2 tablespoons olive oil
1 tablespoon raspberry vinegar
2 tablespoons apple juice

* *Pearl barley has the lowest GI value of the 'true' grains (i.e. excluding the pseudo-grains) at just 25. Its high-soluble fibre content helps slow down the absorption of sugar, lowering the overall GI value of a meal.*

1 Cook the barley in a saucepan of boiling water for 30 minutes or until *al dente*. Refresh under cold running water, then drain well and transfer to a large mixing bowl.

2 Meanwhile, put the onion, peppercorns and salt in a large saucepan with 1 litre (35 fl oz/4 cups) water and bring to the boil over high heat. Add the chicken, reduce the heat to low, cover and simmer gently for 5 minutes. Remove from the heat and leave the chicken in the poaching liquid, covered, for 30 minutes. Remove the chicken and discard the liquid. Set aside to cool, then shred.

3 Using a peeler or mandoline, cut the zucchini and squash into thin ribbons. Place in a heatproof bowl, cover with boiling water and set aside for 1 minute. Refresh under cold running water, then drain well.

4 Add the shredded chicken, zucchini, squash, herbs, pine nuts and currants to the bowl with the barley and stir to combine.

5 To make the dressing, put all the ingredients in a small bowl and whisk to combine. Add to the salad and gently toss to combine. Season with sea salt and freshly ground black pepper, to taste.

TIP Zucchini and squash do not need much cooking, and are delicious raw in fact. Blanching them in boiling water maintains their texture, so they give flavour and a slight crunch to salads.

Barley & mushroom ragout with grilled chicken

PREPARATION TIME: 20 MINUTES
COOKING TIME: 1 HOUR
SERVES 4

2 tablespoons olive oil
1 tablespoon unsalted butter
150 g (5½ oz) Swiss brown mushrooms, sliced
150 g (5½ oz) button mushrooms, sliced
1 large leek, trimmed, halved lengthways and thinly sliced
2 celery stalks, trimmed and diced
2 garlic cloves, crushed
2 teaspoons chopped thyme
200 g (7 oz/1 cup) pearl barley, briefly rinsed
125 ml (4 fl oz/½ cup) white wine
1 litre (35 fl oz/4 cups) chicken or vegetable stock
8 chicken tenderloins
2 tablespoons sour cream (optional)
Snipped chives, to garnish (optional)
Green salad, to serve

Barley works wonderfully in ragout and risotto. If using pre-soaked barley, cook for 20–25 minutes and reduce the stock by 250 ml (9 fl oz/1 cup).

1 Heat half the olive oil and the butter in a large heavy-based saucepan over high heat. Add the mushrooms and cook, stirring occasionally, for 5 minutes or until golden. Transfer to a plate and set aside.

2 Return the pan to medium heat, add the remaining oil and cook the leek and celery, stirring occasionally, for 5–6 minutes or until soft. Add the garlic and thyme and cook, stirring, for 1 minute more.

3 Add the barley to the pan and stir to coat the grains in the oil. Add the wine and simmer until it has reduced by half. Add the stock and bring to the boil, then reduce the heat to low, cover and simmer, stirring occasionally, for 30 minutes. Add the mushrooms and stir to combine, then cover and simmer for a further 10–15 minutes or until the barley is tender and creamy and the stock has been absorbed.

4 Meanwhile, heat a large non-stick frying pan over high heat and spray with olive oil. Cook the tenderloins for 3 minutes each side or until golden and cooked through.

5 Stir the sour cream through the ragout, if desired, and season with sea salt and freshly ground black pepper, to taste. Divide the ragout among 4 serving plates or bowls, top with the chicken and garnish with chives, if desired. Serve immediately with a large green salad.

TIPS Any combination of mushrooms can be used, such as oyster, portobello and so on. Browning the mushrooms first, then adding them towards the end of the cooking time means they won't become too soft.

Serving this ragout with a green salad adds more vegetables and also balances the richness of the dish.

wheat-free

Slow-cooked lamb shank & barley soup with gremolata

PREPARATION TIME: 30 MINUTES
COOKING TIME: 2 HOURS 20 MINUTES
SERVES 4–6

2 tablespoons olive oil
4 small lamb shanks, french trimmed
100 g (3½ oz) sliced pancetta, rind removed, diced
2 carrots, peeled and diced
2 celery stalks, trimmed and diced
1 large brown onion, finely chopped
2 garlic cloves, crushed
2 teaspoons finely chopped rosemary
1 litre (35 fl oz/4 cups) chicken stock
400 g (14 oz) tin chopped tomatoes
70 g (2½ oz/⅓ cup) pearl barley, briefly rinsed
2 fresh bay leaves

GREMOLATA

¼ cup chopped flat-leaf (Italian) parsley
2 teaspoons finely grated lemon zest
1 garlic clove, crushed

1 Heat half the oil in a stockpot or large saucepan over medium–high heat. Add the lamb and cook, turning, for 4–5 minutes or until well browned. Transfer to a plate.

2 Heat the remaining oil in the stockpot. Add the pancetta, carrots, celery and onion and cook, stirring, for 8 minutes or until soft. Add the garlic and rosemary and cook for 1 minute more, then return the lamb to the stockpot.

3 Add the stock, tomatoes, pearl barley and bay leaves to the stockpot. Cover and bring to the boil, then reduce the heat to low and simmer, covered, for 2 hours or until the meat falls from the bone. Use tongs to transfer the lamb to a plate. Remove the meat from the bones and coarsely shred. Skim and discard any excess fat from the surface of the soup, then add the shredded meat and season with sea salt and freshly ground black pepper, to taste.

4 To make the gremolata, combine all the ingredients in a small bowl. Serve the soup sprinkled with gremolata and accompanied by crusty bread.

TIPS This soup can be frozen for up to 2 months. Allow it to cool, then divide among containers, cover with airtight lids and freeze.

If you leave the soup to cool in the refrigerator overnight before shredding the lamb, the fat will solidify on top of the soup and it will be easier to remove.

** Pearl barley makes a great addition to winter soups, adding flavour and texture as well as dietary fibre. The low-GI content of barley means it will release energy slowly, leaving you feeling fuller for longer.*

Barley, tofu & spinach salad with miso dressing

PREPARATION TIME: 20 MINUTES
COOKING TIME: 30 MINUTES
SERVES 4

1 tablespoon olive oil

2 teaspoons finely grated ginger

1 garlic clove, crushed

200 g (7 oz/1 cup) pearl barley, briefly rinsed

500 ml (17 fl oz/2 cups) chicken or vegetable stock

1 bunch asparagus, trimmed, halved lengthways and cut into 3 cm (1¼ inch) lengths

200 g (7 oz) marinated tofu, sliced

1 small firm ripe avocado, halved, stone removed, peeled and sliced (see tips)

100 g (3½ oz) baby spinach leaves

1 tablespoon sesame seeds, lightly toasted

MISO DRESSING

2 tablespoons white or yellow (brown) miso paste

2 tablespoons mirin

1½ tablespoons tamari

1 tablespoon rice wine vinegar

½ teaspoon caster (superfine) sugar

* *This salad makes a balanced meal for vegetarians as it combines barley with tofu and sesame seeds, which are both great sources of protein.*

1 Heat the olive oil in a large saucepan over medium heat, add the ginger and garlic and cook, stirring, for 30 seconds. Add the barley and cook, stirring, for 2 minutes, then add the stock and 500 ml (17 fl oz/2 cups) water and bring to the boil. Reduce the heat to low and simmer for 25 minutes or until *al dente*. Rinse under cold running water, then drain well and transfer to a large mixing bowl.

2 Meanwhile, blanch the asparagus in a saucepan of boiling water until bright green and tender-crisp. Refresh under cold running water, then drain well.

3 To make the miso dressing, put all the ingredients in a small bowl and stir to combine.

4 Add the asparagus, tofu, avocado, baby spinach and half the dressing to the barley and gently toss to combine. Divide the salad among serving bowls, drizzle with a little of the remaining dressing and sprinkle with the sesame seeds.

TIPS Squeeze a little lemon juice over the sliced avocado to prevent it discolouring.

Pearl barley works beautifully with the earthy flavours of miso and tamari. Toasting the barley before cooking, and adding ginger and garlic to the cooking liquid, give lots of additional flavour.

PORK, TOMATO, PAPRIKA
& BARLEY STEW (PAGE 194)

Pork, tomato, paprika & barley stew

PREPARATION TIME: 20 MINUTES
COOKING TIME: 2 HOURS 15 MINUTES
SERVES 4

2 tablespoons olive oil
800 g (1 lb 12 oz) trimmed and diced pork shoulder
1 large red onion, diced
2 celery stalks, diced
100 g (3½ oz) speck (see tips), diced
2 garlic cloves, thinly sliced
2 teaspoons hot smoked paprika
2 thyme sprigs
60 ml (2 fl oz/¼ cup) Spanish sherry
1 tablespoon tomato paste (concentrated purée)
400 g (14 oz) tin chopped tomatoes
70 g (2½ oz/⅓ cup) pearl barley, briefly rinsed
250 ml (9 fl oz/1 cup) chicken stock
1 teaspoon honey
Rocket (arugula) leaves and crusty bread, to serve

GREEN OLIVE RELISH

90 g (3¼ oz/½ cup) green olives, pitted and sliced
2 tablespoons chopped flat-leaf (Italian) parsley

** I love adding pearl barley to stews such as this, as it provides body and texture and reduces the amount of meat needed. Serve it with bread to mop up the sauce.*

1 Preheat the oven to 160°C (315°F/Gas 2–3). Heat 1 tablespoon of the oil in a large flameproof casserole dish over high heat and cook the pork, in batches, stirring, for 3–4 minutes or until browned. Transfer to a bowl.

2 Add the remaining oil to the casserole dish and cook the onion, celery and speck, stirring, for 5 minutes or until softened. Add the garlic, paprika and thyme and cook, stirring, for 1 minute more. Add the sherry and simmer until it has almost evaporated, then add the tomato paste and cook, stirring, for 1 minute. Add the tomatoes, barley and stock and stir to combine, then bring to the boil.

3 Return the pork to the casserole dish and stir to coat in the tomato mixture. Cover with a lid or foil and bake for 2 hours or until the pork is very tender. Stir through the honey and season with sea salt and freshly ground black pepper, to taste.

4 Meanwhile, to make the green olive relish, combine the olives and parsley in a small bowl.

5 Divide the stew among serving plates, sprinkle with the relish and serve with rocket leaves and crusty bread.

TIPS You can substitute bacon or pancetta for the speck.
This stew can be frozen for up to 2 months. Allow it to cool, then divide among airtight containers, cover with the lids and freeze.

Barley & oat porridge with fig & hazelnuts

PREPARATION TIME: 5 MINUTES
COOKING TIME: 10 MINUTES
SERVES 4

95 g (3¼ oz/1 cup) rolled (porridge) oats

100 g (3½ oz/1 cup) rolled barley

45 g (1¾ oz/¼ cup) chopped dried dessert figs

500 ml (17 fl oz/2 cups) milk

50 g (1¾ oz/⅓ cup) hazelnuts, lightly toasted, skinned and chopped

Honey, to drizzle

* *Rolled barley makes a deliciously nutty porridge. I find using equal quantities of barley and oats gives the best result. Barley is packed with beta-glucan soluble fibre and is therefore linked to reduced risk of heart disease.*

1 Put the oats, barley, figs, milk and 375 ml (13 fl oz/1½ cups) water in a large saucepan. Bring to the boil over high heat, stirring occasionally, then reduce the heat to low and simmer, stirring, for 5 minutes or until thick and creamy.

2 Divide the porridge among 4 serving bowls, top with chopped hazelnuts and a drizzle of honey, and serve.

TIP You can use any combination of dried fruit and nuts in this porridge, such as dried apricots, currants, dates, walnuts, almonds and/or pistachios.

Baked barley puddings with port-poached prunes

PREPARATION TIME: 15 MINUTES
COOKING TIME: 30 MINUTES
SERVES 4

160 g (5½ oz/1 cup) cooked pearl barley (see page 183)
2 eggs
375 ml (13 fl oz/1½ cups) milk
2 tablespoons caster (superfine) sugar
1 teaspoon natural vanilla extract
½ teaspoon ground nutmeg, to sprinkle

POACHED PRUNES

100 g (3½ oz) pitted prunes, quartered
2 tablespoons light brown sugar
2 tablespoons port

1 Preheat the oven to 160°C (315°F/Gas 2–3). Lightly grease four 150 ml (5 fl oz) ovenproof ramekins and place on a large baking tray.

2 Divide the barley evenly among the greased ramekins. Whisk the eggs, milk, sugar and vanilla until well combined, then pour evenly over the barley. Sprinkle the top of each pudding with a little nutmeg and bake for 30 minutes or until puffed and just set (the puddings should have a slight wobble in the centre). Set aside to cool slightly.

3 Meanwhile, to make the poached prunes, put all the ingredients in a medium saucepan with 125 ml (4 fl oz/½ cup) water and cook over medium heat, stirring to dissolve the sugar. Bring to the boil, then reduce the heat to low and simmer for 5 minutes or until the prunes are syrupy. Remove from the heat and allow to cool a little.

4 Serve the puddings topped with the poached prunes.

TIP These puddings can be served warm or chilled.

Pearl barley can be used as a substitute for short-grain rice in baked puddings. I find cooking the pearl barley first and then baking it in the oven is easier and faster than cooking it on the stovetop.

In comparison to white-rice pudding, these barley puddings are higher in fibre and protein.

Lemon barley water

PREPARATION TIME: 5 MINUTES
COOKING TIME: 15 MINUTES
MAKES 1.25 LITRES (44 FL OZ/5 CUPS)

100 g (3½ oz/½ cup) pearl barley

2 lemons, plus lemon slices, extra, to serve

110 g (3¾ oz/½ cup) caster (superfine) sugar

** Barley water is an ancient home remedy that was given to the sick for ailments such as stomach upsets, cystitis, coughs and colds, to name but a few.*

1 Put the barley in a sieve and rinse under cold running water until the water runs clear. Use a vegetable peeler to remove the lemon zest from the lemons, then juice the lemons.

2 Put the barley, lemon zest and 1.5 litres (52 fl oz/6 cups) water in a large saucepan and bring to the boil, then reduce the heat to low and simmer for 10 minutes. Strain, discarding the barley and zest.

3 Add the sugar to the barley liquid and stir to dissolve, then stir through the lemon juice. Pour the barley water into a bottle/bottles, seal and chill overnight. Serve with lemon slices.

TIPS Although it is traditionally served undiluted, barley water makes a beautifully refreshing drink topped with ice, lemon slices and sparkling mineral water. It is also delicious served with vodka or gin for a summer cocktail.

Keep in the refrigerator for up to 1 week.

Roasted nectarines with barley & almond crumble

PREPARATION TIME: 15 MINUTES
COOKING TIME: 20 MINUTES
SERVES 4–6

6 firm ripe slipstone (freestone) nectarines, halved and stones removed

50 g (1¾ oz/½ cup) rolled barley

75 g (2¾ oz/⅓ cup, firmly packed) light brown sugar

35 g (1¼ oz/¼ cup) plain (all-purpose) flour

1 teaspoon ground cinnamon

35 g (1¼ oz/⅓ cup) flaked almonds

60 g (2¼ oz) unsalted butter, chilled and diced

Vanilla ice cream, plain yoghurt or cream, to serve

** Rolled barley can be used in recipes where you would typically used rolled (porridge) oats, such as here. Rolled barley tends to be slightly firmer than rolled oats, so it makes a lovely crisp topping.*

1 Preheat the oven to 180°C (350°F/Gas 4). Lightly grease a 1 litre (35 fl oz/4 cup) ovenproof baking dish. Place the nectarines, cut side up, in a single layer in the greased dish.

2 Put the barley, sugar, flour, cinnamon and almonds in a medium bowl. Use clean fingertips to rub in the butter until small clumps form. Sprinkle the crumble mixture evenly over the nectarines.

3 Roast the nectarines for 20 minutes or until the crumble is golden and the nectarines are tender. Serve with vanilla ice cream, plain yoghurt or cream.

TIP You can replace the nectarines with other seasonal stone fruit, such as plums or peaches.

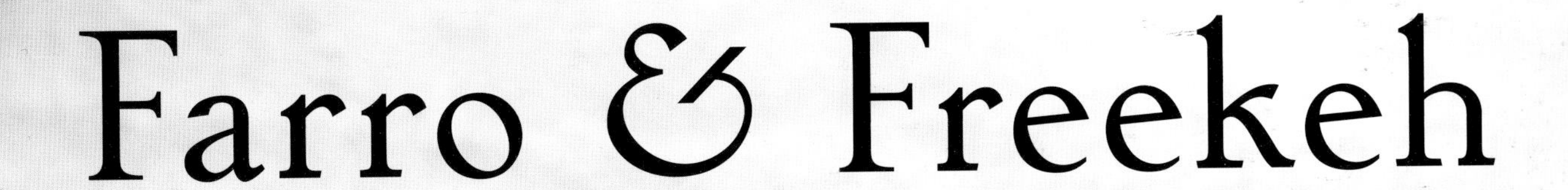

Farro & Freekeh

NUTRITIONAL INFORMATION
(per 85 g/3 oz/½ cup uncooked cracked farro)

Energy: 1397 kilojoules, 332 calories **Protein:** 12.9 g

Total fat: 1.9 g **Saturated fat:** <1 g **Carbohydrate:** 52.2 g

Sugars: <1 g **Dietary fibre:** 4.6 g **Sodium:** <1 g

NUTRITIONAL INFORMATION
(per 100 g/3½ oz/½ cup uncooked whole-grain freekeh)

Energy: 1471 kilojoules, 350 calories **Protein:** 12.6 g **Total fat:** 2.7 g

Carbohydrate: 72 g **Dietary fibre:** up to 16.5 g **Iron:** 4.5 mg

Calcium: 53 mg **Magnesium:** 110 mg **Sodium:** 6 mg

Farro (emmer wheat) is an ancient grain, thought to pre-date all others, that fed the people of the Near East and Mediterranean for thousands of years. Ground into a paste and cooked, it was the primary ingredient in puls, the polenta eaten by the Roman poor. It sustained Roman legions and gave rise to the Italian name of flour.

Today farro is mainly grown as a speciality crop in Italy and a few surrounding countries, but its popularity extends well beyond those borders as health-conscious cooks are discovering its appeal.

Farro is rich in fibre, protein, magnesium, niacin and zinc, as well as phytonutrients and antioxidants including lignans, phenolic acids, phytic acid, plant sterols and saponins. As with most varieties of wheat, farro/emmer is unsuitable for sufferers of wheat allergies or coeliac disease. But it is gentler on the digestive system than common wheat as the gluten molecules are not as strong.

Farro is sold as the whole grain or cracked farro. Whole-grain farro takes 50–60 minutes to cook, so is often soaked overnight prior to cooking to reduce this time to 25–30 minutes. Cracked farro (traditionally known as *farro spezzato*) is where the farro grains have been split during the milling process — it requires less cooking time (about 20 minutes) and no pre-soaking.

Farro is delicious added to stews, casseroles, hearty soups such as minestrone, or used as a salad grain. It can also be ground to make flour and can then be used in pasta and bread making. Try using farro in recipes that traditionally use pearl barley or spelt.

There is often confusion between farro (emmer) and spelt, and while closely related and similar in many features, they are not always interchangeable in recipes.

To cook farro

WHOLE GRAIN Add 200 g (7 oz/1 cup) farro to 1.5 litres (52 fl oz/6 cups) boiling water. Return to the boil, then reduce the heat, cover and simmer for 50 minutes or until the grains are *al dente*. Rinse under cold running water, then drain well. If farro has been pre-soaked, reduce the cooking time to 20–30 minutes.

CRACKED Add 175 g (6 oz/1 cup) cracked farro to 1.5 litres (52 fl oz/6 cups) boiling water. Return to the boil, then reduce the heat, cover and simmer for 15–20 minutes or until grains are *al dente*. Rinse under cold running water, then drain well.

Freekeh (pronounced free-ka) is wheat that is picked while green and then put through a roasting process. It dates as far back as the 1200s and is popular in Middle Eastern and North African cuisines.

Freekeh is becoming more readily available, as the nutritional benefits of green wheat have been discovered. Compared to regular wheat, freekeh contains higher levels of protein and dietary fibre, as well as various vitamins and minerals including calcium, potassium, iron and zinc. Whole-grain freekeh has a GI of 43, so it's a great food for diabetics and those on a low-GI diet.

It is available in its whole-grain form or as cracked freekeh, where the grains are cracked during the milling process. Both whole-grain and cracked freekeh are delicious in salads such as tabouleh, and work beautifully with Middle Eastern flavours and spices. They can also be added to soups and stews or used as stuffings.

To cook freekeh

WHOLE GRAIN Add 200 g (7 oz/1 cup) freekeh to 1.5–2 litres (52–70 fl oz/6–8 cups) boiling water. Return to the boil, then reduce the heat, cover and simmer for 45–50 minutes or until the grains are *al dente*. Rinse under cold running water, then drain well.

CRACKED Add 175 g (6 oz/1 cup) cracked freekeh to 1.5–2 litres (52–70 fl oz/6–8 cups) boiling water. Return to the boil, then reduce the heat, cover and simmer for 15 minutes or until the grains are *al dente*. Rinse under cold running water, then drain well.

Farro minestrone

PREPARATION TIME: 20 MINUTES
COOKING TIME: 1 HOUR
SERVES 4

2 tablespoons olive oil

100 g (3½ oz) pancetta, rind removed, diced

1 large onion, finely chopped

2 carrots, peeled and diced

2 celery stalks, trimmed and diced

1 zucchini (courgette), trimmed and diced

3 garlic cloves, crushed

1 teaspoon thyme leaves

2 tablespoons tomato paste (concentrated purée)

5 cm (2 inch) piece parmesan cheese rind

130 g (4½ oz/¾ cup) cracked farro

400 g (14 oz) tin chopped tomatoes

1.25 litres (44 fl oz/5 cups) chicken stock

250 g (9 oz) cavolo nero, trimmed, shredded (see tips)

Grated parmesan cheese, chopped parsley and crusty bread, to serve

** Farro is similar to pearl barley once cooked, so it works well in soups. Compared to regular wheat, it is higher in fibre, protein, zinc and B vitamins.*

Cavolo nero is Tuscan cabbage. Substitute silverbeet (Swiss chard) or English spinach if unavailable.

1 Heat the oil in a large saucepan over medium heat. Add the pancetta, onion, carrots, celery and zucchini and cook, stirring occasionally, for 7–8 minutes or until the vegetables are soft. Add the garlic and thyme and cook, stirring, for 1 minute. Add the tomato paste and cook, stirring, for 1 minute more.

2 Stir in the parmesan rind, farro, tomatoes and stock and bring to the boil. Reduce the heat to low and simmer, stirring occasionally, for 45 minutes or until the soup has thickened. Add the cavolo nero and simmer for a further 5 minutes.

3 Season with sea salt and freshly ground black pepper, to taste. Add a little extra water if the soup has become too thick. Remove the parmesan rind. Ladle the soup into serving bowls, sprinkle with the grated parmesan and parsley and serve with crusty bread.

TIPS Remove and discard the stalks from the cavolo nero before shredding the leaves.

This soup can be frozen for up to 2 months. Allow to cool, then divide among airtight containers, cover with the lids and freeze.

Roasted capsicum & freekeh salad with labne

PREPARATION TIME: 20 MINUTES
COOKING TIME: 45 MINUTES
SERVES 4

200 g (7 oz/1 cup) freekeh

2 large red capsicums (peppers), halved and seeded

400 g (14 oz) tin chickpeas, drained and rinsed

2 tablespoons snipped chives

75 g (2¾ oz) baby rocket (arugula) leaves

120 g (4¼ oz) labne (see tip)

DRESSING

1½ tablespoons olive oil

1½ tablespoons lemon juice

1 teaspoon sumac, plus extra, to serve

2 teaspoons honey

* *Freekeh is wheat that is picked while young and the grain is green. It is higher in nutrients such as protein, fibre, calcium, iron and zinc, than mature wheat. Freekeh has over five times as much fibre and twice as much protein as white rice.*

Both whole-grain and cracked freekeh are ideal to use in salads where you would typically use rice or couscous.

1 Cook the freekeh in a large saucepan of lightly salted boiling water for 45 minutes or until tender. Rinse under cold running water, then drain well.

2 Meanwhile, preheat the grill (broiler) on high. Place the capsicums, skin side up, on a baking tray and grill until the skin is blistered, about 6–8 minutes. Transfer to a heatproof bowl and cover with plastic wrap. Set aside to cool slightly, then carefully peel away and discard the skin and thinly slice the flesh. Set aside.

3 To make the dressing, whisk together the olive oil, lemon juice, sumac and honey.

4 Put the cooked freekeh, capsicums, chickpeas, chives and rocket in a large bowl. Add the dressing and gently toss to combine, then season with salt and pepper, to taste. Serve topped with the labne and sprinkled with a little extra sumac.

TIP Labne is yoghurt cheese made from strained plain yoghurt. It has a tangy taste and creamy texture. If you are unable to find it you can substitute marinated feta cheese.

Oxtail & farro ragù

PREPARATION TIME: 30 MINUTES
COOKING TIME: 3 HOURS 30 MINUTES
SERVES 4–6

35 g (1¼ oz/¼ cup) plain (all-purpose) flour
1.25 kg (2 lb 12 oz) oxtail pieces
2 tablespoons olive oil
2 large onions, finely chopped
2 large carrots, peeled and diced
2 celery stalks, trimmed and diced
125 g (4½ oz) shortcut bacon, rind removed and diced
3 garlic cloves, thinly sliced
3 anchovy fillets, chopped
2 teaspoons chopped rosemary
2 tablespoons tomato paste (concentrated purée)
500 ml (17 fl oz/2 cups) red wine
500 ml (17 fl oz/2 cups) beef stock
90 g (3¼ oz/½ cup) cracked farro
Wholemeal (whole-wheat) spaghetti, to serve
Finely grated parmesan cheese, to garnish

* *The farro's role in this ragù is threefold: it gives texture to complement the softness of the oxtail, it thickens the braising liquid and it helps cut through the richness of the meat. Farro makes a good addition to soups and stews, and is rich in protein, fibre, zinc and B vitamins.*

1 Preheat the oven to 160°C (315°F/Gas 2–3). Place the flour on a large plate and season with sea salt and freshly ground black pepper. Toss the oxtail in the seasoned flour, shaking off any excess. Heat half the oil in a large flameproof casserole dish and brown the oxtail, in batches, for 2–3 minutes. Transfer to a bowl and set aside.

2 Return the dish to medium heat and add the remaining oil, the onions, carrots, celery and bacon. Cook, stirring occasionally, for 6–7 minutes or until the vegetables have softened. Add the garlic, anchovies and rosemary and cook, stirring, for 1 minute more.

3 Add the tomato paste and cook, stirring, for 1 minute. Add the wine and simmer for 2 minutes, then add the stock and 250 ml (9 fl oz/1 cup) water and bring to the boil. Return the oxtail to the dish, cover and bake for 2 hours. Stir in the farro, then cover and bake for a further 1 hour or until the meat is tender and falling off the bone. Remove from the oven and set aside to cool slightly.

4 Use tongs to transfer the oxtail to a plate. Remove the meat from the bones and coarsely shred. Skim any excess fat from the surface of the ragù. Return the shredded meat to the casserole dish and season with sea salt and freshly ground black pepper, to taste.

5 Serve the ragù tossed with spaghetti and garnished with parmesan.

TIP You could also serve this ragù with mashed potato or soft polenta. I also like to serve it with a salad, to make it a more complete meal and balance the richness of the ragù.

SLOW-ROASTED LAMB SALAD WITH FREEKEH & POMEGRANATE (PAGE 214)

Slow-roasted lamb salad with freekeh & pomegranate

PREPARATION TIME: 25 MINUTES
COOKING TIME: 3 HOURS–3 HOURS 30 MINUTES
SERVES 4

1 teaspoon cumin seeds
1/2 teaspoon ground cinnamon
1.25 kg (2 lb 12 oz) boneless lamb shoulder
1 tablespoon olive oil, plus 1 1/2 tablespoons, extra
60 ml (2 fl oz/1/4 cup) pomegranate molasses (see tip), plus 1 tablespoon, extra
1 tablespoon honey
175 g (6 oz/1 cup) cracked freekeh
2 bunches broccolini, trimmed and cut into long florets
1/2 small red onion, thinly sliced
1 pomegranate, seeds removed and juice reserved

** You can use whole-grain freekeh in this recipe, but it will take 45 minutes to cook.*

Unlike mature wheat, freekeh is rich in lutein and zeaxanthin, which are linked to the prevention of macular degeneration.

1 Preheat the oven to 160°C (315°F/Gas 2–3). Crush the cumin seeds and cinnamon in a mortar and pestle. Drizzle the lamb with the oil and rub with the crushed spices, then season well with sea salt and freshly ground black pepper. Brown the lamb in a large non-stick frying pan over medium–high heat, turning to brown both sides.

2 Place the lamb shoulder in a large roasting pan and brush with about three-quarters of the combined pomegranate molasses and honey. Place a large piece of non-stick baking paper over the lamb, tucking the edges under, then cover the entire pan with a layer of foil. Roast the lamb for 2 1/2–3 hours or until the meat is extremely tender and can be shredded with a fork. Brush the remaining pomegranate mixture over the lamb, then set aside to cool slightly. Coarsely shred the meat.

3 Meanwhile, cook the freekeh in a large saucepan of lightly salted boiling water for 15–20 minutes or until *al dente*. Rinse under cold running water, then drain well, squeezing out as much excess water as possible. Transfer to a large bowl.

4 Blanch the broccolini in a large saucepan of boiling water until bright green and tender-crisp. Refresh under cold running water, then drain well. Add the broccolini, onion and three-quarters of the pomegranate seeds to the freekeh and stir to combine.

5 Whisk together the extra olive oil and pomegranate molasses and 1 tablespoon of the reserved pomegranate juice. Add to the salad and toss to combine, then season with salt and pepper, to taste.

6 To serve, pile the salad onto a large serving platter, top with the shredded lamb and sprinkle with the remaining pomegranate seeds.

TIP Pomegranate molasses is made from reduced pomegranate juice and has a rich sweet-and-sour taste, similar to tamarind. It is available from speciality food stores and Middle Eastern delicatessens. You can substitute 2 teaspoons tamarind pulp or 2 tablespoons lemon juice.

Red wine farro risotto

PREPARATION TIME: 15 MINUTES
COOKING TIME: 50 MINUTES
SERVES 4

1 litre (35 fl oz/4 cups) vegetable or chicken stock

1 tablespoon olive oil

1 tablespoon butter

1 large red onion, finely chopped

2 garlic cloves, crushed

260 g ($9\frac{1}{4}$ oz/$1\frac{1}{2}$ cups) cracked farro

375 ml (13 fl oz/$1\frac{1}{2}$ cups) red wine

50 g ($1\frac{3}{4}$ oz/$\frac{1}{2}$ cup) finely grated parmesan cheese, plus extra, to serve

Chopped flat-leaf (Italian) parsley, to serve

* *Farro produces a risotto with a slightly nutty flavour and chewy texture, similar to pearl barley. The earthiness of the farro is well matched with red wine and parmesan here.*

This risotto is quite rich and very filling, so you only need to serve small portions. To make a balanced meal, serve it with a salad of bitter greens, such as witlof, rocket (arugula) and thinly sliced pear.

1 Bring the stock to the boil in a medium saucepan over high heat, then reduce the heat to very low and keep at a gentle simmer.

2 Heat the olive oil and butter in a large heavy-based saucepan over medium heat. Add the onion and cook, stirring occasionally, for 5 minutes or until soft. Add the garlic and cook, stirring, for 1 minute.

3 Add the farro and cook, stirring, for 1–2 minutes or until evenly coated in the oil mixture. Add 250 ml (9 fl oz/1 cup) of the wine and let it bubble away until reduced by half. Start adding the stock, a ladleful at a time, and cook, stirring, until it is absorbed before adding more. When about half the stock is used, add the remaining wine, then continue adding the remaining stock a ladleful at a time.

4 When you have added all the stock and the farro is tender, about 30–35 minutes, remove the risotto from the heat and add the parmesan. Cover and set aside for 5 minutes. Serve immediately, garnished with parsley and extra parmesan.

TIP For a really indulgent risotto, add a knob of butter with the parmesan at the end.

Chicken & freekeh tagine with lemon and green olives

PREPARATION TIME: 30 MINUTES
(+ 1 HOUR MARINATING)
COOKING TIME: 1 HOUR 30 MINUTES
SERVES 4

1/3 cup firmly packed coriander (cilantro) leaves
1/3 cup firmly packed parsley leaves
2 garlic cloves, crushed
1 long red chilli, seeded and finely chopped
2 teaspoons ground cumin
1 teaspoon ground turmeric
2 tablespoons lemon juice
2 tablespoons olive oil
1.5 kg (3 lb 5 oz) chicken pieces on the bone (we used thighs and legs)
1 large red onion, thinly sliced
4 strips lemon zest
500 ml (17 fl oz/2 cups) chicken stock
100 g (3½ oz/½ cup) freekeh
50 g (1¾ oz/¼ cup) pitted dried dates, chopped
90 g (3¼ oz/½ cup) green olives
Rocket (arugula) leaves, to garnish

1 Put the coriander, parsley, garlic, chilli, cumin, turmeric, lemon juice and half the olive oil in a blender or food processor and blend to a smooth paste, adding 1–2 tablespoons hot water if necessary. Season with sea salt and freshly ground black pepper.

2 Place the chicken pieces in a large shallow glass or ceramic container, add half the marinade (reserve the remaining marinade) and turn to coat. Cover and refrigerate for at least 1 hour.

3 Preheat the oven to 180°C (350°F/Gas 4). Heat the remaining oil in a large flameproof casserole dish over high heat. Brown the chicken, in batches, for 2–3 minutes or until golden. Transfer to a bowl and set aside.

4 Return the dish to medium heat, add the onion and cook, stirring occasionally, for 5 minutes or until softened. Add the reserved marinade and cook, stirring, for 1–2 minutes or until fragrant. Return the chicken to the dish with the lemon zest, stock, freekeh and dates and bring to the boil. Cover and bake for 1 hour. Add the olives and bake, covered, for a further 15 minutes or until the freekeh is tender. Skim any fat from the surface. Serve garnished with extra rocket leaves.

TIP You can use a whole chicken, cut into 8 pieces, for this recipe.

* *Freekeh works really well in slow-cooked dishes, such as this tagine, and soups. Its herbaceous flavour is delicious with the freshness of lemon, coriander and parsley. Whole-grain freekeh has a GI of just 43.*

Warm farro, pancetta & parsley salad

PREPARATION TIME: 15 MINUTES
COOKING TIME: 30 MINUTES
SERVES 4

220 g (7¾ oz/1¼ cups) cracked farro
125 g (4½ oz) pancetta, rind removed, cut into lardons
1 French shallot, finely chopped
2 celery stalks, trimmed and cut into thin batons
2 tablespoons finely chopped flat-leaf (Italian) parsley

RED WINE VINAIGRETTE
2 tablespoons olive oil
2 teaspoons red wine vinegar
1 tablespoon lemon juice
1 garlic clove, crushed
½ teaspoon honey

* *Farro can be purchased as either whole-grain or cracked. The most readily available is cracked farro and it cooks in about half the time whole-grain requires. If using whole-grain farro, soak it in cold water for several hours before cooking, then simmer in boiling water for up to 1 hour, until* al dente.

1 Cook the farro in a large saucepan of lightly salted boiling water for 15–20 minutes or until *al dente*. Rinse briefly under cold running water, then drain well and transfer to a large bowl.

2 To make the red wine vinaigrette, whisk all the ingredients together in a small bowl.

3 Heat a large, deep frying pan over high heat and cook the pancetta, stirring, for 3–4 minutes or until golden. Reduce the heat to medium and add the farro, red wine vinaigrette, shallot, celery and parsley. Season with sea salt and freshly ground black pepper and stir to combine. Cook until heated through, then serve.

TIP This warm salad is delicious served with grilled (broiled) meat, such as chicken or pork.

Index

D

F

G

Q

R

S

T

U

V

W

Whole grain suppliers

Olive Green Organics
www.olivegreenorganics.com.au

Four Leaf Milling
www.fourleafmilling.com.au

Ceres Organics
www.ceres.co.nz

The Chia Co
www.thechiaco.com.au

Mount Zero
www.mountzeroolives.com

Eden Organic
www.edenfoods.com

Bob's Red Mill
www.bobsredmill.com

Bluebird Grain Farms
www.bluebirdgrainfarms.com

Published in 2013 by Murdoch Books Pty Limited

Murdoch Books Australia
83 Alexander St
Crows Nest, NSW 2065
Australia
Phone: +61 (0) 2 8425 0100
Fax: +61 (0) 2 9906 2218
www.murdochbooks.com.au
info@murdochbooks.com.au

Murdoch Books UK Limited
Erico House, 6th Floor
93–99 Upper Richmond Road
Putney, London SW15 2TG
Phone: +44 (0) 20 8785 5995
Fax: +44 (0) 20 8785 5985
www.murdochbooks.co.uk
info@murdochbooks.co.uk

For Corporate Orders & Custom Publishing contact
Noel Hammond, National Business Development
Manager Murdoch Books Australia

Publisher: Anneka Manning
Recipes, text and styling: Chrissy Freer
Designer: Susanne Geppert
Photographer: Julie Renouf
Project Editor: Martina Vascotto
Editor: Anna Scobie
Food Editor: Christine Osmond
Home Economist: Caroline Griffiths
Production: Alexandra Gonzalez

A cataloguing-in-publication entry is available from the catalogue of the National Library of Australia at www.nla.gov.au.

A catalogue record for this book is available from the British Library.

Printed by Hang Tai Printing Company Limited, China

The Publisher and stylist would like to thank Breville, Country Road, Est Australia, Market Import, Mozi, Mud Australia, Royal Hamam, Status Collections, The Works, Bed Bath N' Table and Urban Edge Ceramics for lending equipment and props for use and photography.

IMPORTANT: Those who might be at risk from the effects of salmonella poisoning (the elderly, pregnant women, young children and those suffering from immune deficiency diseases) should consult their doctor with any concerns about eating raw eggs.

OVEN GUIDE: You may find cooking times vary depending on the oven you are using. For fan-forced ovens, as a general rule, set the oven temperature to 20°C (35°F) lower than indicated in the recipe.

MEASURES: We have used 20 ml (4 teaspoon) tablespoon measures. If you are using a 15 ml (3 teaspoon) tablespoon add an extra teaspoon of the ingredient for each tablespoon specified.